श्री BalaJi

INORGANIC CHEMISTRY CHAPTERWISE

15 YEARS (2011-2025)

PAPERS with Detailed Solutions

JEE

(MAIN & ADVANCED)

SHRI BALAJI PUBLICATIONS
(EDUCATIONAL PUBLISHERS & DISTRIBUTORS)
[AN ISO 9001-2008 CERTIFIED ORGANIZATION]
Muzaffarnagar (U.P.) - 251001

Published by:

SHRI BALAJI PUBLICATIONS

(EDUCATIONAL PUBLISHERS & DISTRIBUTORS)

[AN ISO 9001-2008 CERTIFIED ORGANIZATION]

6, Gulshan Vihar,
Jansath Road, Muzaffarnagar (U.P.)
Phone : 0131-2660440 (O)
website : www.shribalajibooks.com
email : sbjpub@gmail.com

Edition : **2026**

Price : ₹ 250.00

CONTENTS

GROW GREEN

Save Nature

CHAPTER

1

Classification of Elements and Periodic Properties

A. Objective Questions

[Only one option is correct]

1. The first ionisation potential of Na is 5.1 eV. The value of electron gain enthalpy of Na^+ will be: **[JEE (Mains) 2013]**

(a) –2.55 eV (b) –5.1 eV (c) –10.2 eV (d) +2.55 eV

2. The ionic radii (in Å) of N^{3-}, O^{2-} and F^- are respectively : **[JEE (Mains) 2015]**

(a) 1.71, 1.40 and 1.36 (b) 1.71, 1.36 and 1.40
(c) 1.36, 1.40 and 1.71 (d) 1.36, 1.71 and 1.40

3. Which of the following atoms has the highest first ionization energy ? **[JEE (Mains) 2016]**

(a) Rb (b) Na (c) K (d) Sc

4. Both lithium and magnesium display several similar properties due to the diagonal relationship, however, the one which is incorrect, is: **[JEE (Mains) 2017]**

(a) both form nitrides
(b) nitrates of both Li and Mg yield NO_2 and O_2 on heating
(c) both form basic carbonates
(d) both form soluble bicarbonates

5. The IUPAC symbol for the element with atomic number 119 would be : **[JEE (Mains) 2019]**

(a) une (b) unh (c) uun (d) uue

6. The elements with atomic numbers 101 and 104 belong to, respectively: **[JEE Main (Sept.) 2020]**

(a) Actinoids and Group 6 (b) Actinoids and Group 4
(c) Group 11 and Group 4 (d) Group 6 and Actinoids

7. The set that contains atomic numbers of only transition elements, is: **[JEE Main (Sept.) 2020]**

(a) 37, 42, 50, 64 (b) 21, 25, 42, 72 (c) 9, 17, 34, 38 (d) 21, 32, 53, 64

8. The atomic radius of Ag is closest to: **[JEE Main (Jan.) 2020]**

(a) Ni (b) Cu (c) Au (d) Hg

9. The increasing order of the atomic radii of the following elements is: **[JEE Main (Jan.) 2020]**

(A) C (B) O
(C) F (D) Cl (E) Br

(a) (A) < (B) < (C) < (D) < (E) (b) (D) < (C) < (B) < (A) < (E)
(c) (C) < (B) < (A) < (D) < (E) (d) (B) < (C) < (D) < (A) < (E)

10. The ionic radii of O^{2-}, F^-, Na^+ and Mg^{2+} are in the order: **[JEE Main (Sept.) 2020]**

(a) $O^{2-} > F^- > Mg^{2+} > Na^+$ (b) $Mg^{2+} > Na^+ > F^- > O^{2-}$
(c) $F^- > O^{2-} > Na^+ > Mg^{2+}$ (d) $O^{2-} > F^- > Na^+ > Mg^{2+}$

11. The correct order of the ionic radii of O^{2-}, N^{3-}, F^-, Mg^{2+}, Na^+ and Al^{3+} is: **[JEE Main (Sept.) 2020]**

(a) $N^{3-} < O^{2-} < F^- < Na^+ < Mg^{2+} < Al^{3+}$
(b) $Al^{3+} < Mg^{2+} < Na^+ < F^- < O^{2-} < N^{3-}$
(c) $Al^{3+} < Na^+ < Mg^{2+} < O^{2-} < F^- < N^{3-}$
(d) $N^{3-} < F^- < O^{2-} < Mg^{2+} < Na^+ < Al^{3+}$

12. The third ionization enthalpy is minimum for : **[JEE Main (Jan.) 2020]**

(a) Ni (b) Co (c) Mn (d) Fe

13. The first ionization energy (in kJ/mol) of Na, Mg, Al and Si respectively, are : **[JEE Main (Jan.) 2020]**

(a) 496, 577, 786, 737 (b) 786, 737, 577, 496
(c) 496, 577, 737, 786 (d) 496, 737, 577, 786

14. B has a smaller first ionization enthalpy than Be. Consider the following statements:

(i) it is easier to remove $2p$ electron than $2s$ electron.
(ii) $2p$ electron of B is more shielded from the nucleus by the inner core of electrons than the $2s$ electrons of Be.
(iii) $2s$ electron has more penetration power than $2p$ electron.
(iv) atomic radius of B is more than Be (atomic number B = 5, Be = 4).

The correct statements are: **[JEE Main (Jan.) 2020]**

(a) (i), (iii) and (iv) (b) (i), (ii) and (iii)
(c) (i), (ii) and (iv) (d) (ii), (iii) and (iv)

15. The first and second ionisation enthalpies of a metal are 496 and 4560 kJ mol^{-1}, respectively. How many moles of HCl and H_2SO_4, respectively, will be needed to react completely with 1 mole of the metal hydroxide? **[JEE Main (Jan.) 2020]**

(a) 1 and 1 (b) 1 and 2 (c) 2 and 0.5 (d) 1 and 0.5

16. The five successive ionization enthalpies of an element are 800, 2427, 3658, 25024 and 32824 kJ mol^{-1}. The number of valence electrons in the element is: **[JEE Main (Sept.) 2020]**

(a) 4 (b) 2 (c) 5 (d) 3

17. In general, the property (magnitudes only) that shows an opposite trend in comparison to other properties across a period is : **[JEE Main (Sept.) 2020]**

(a) Ionization enthalpy (b) Electron gain enthalpy
(c) Atomic radius (d) Electronegativity

18. Among the statements (I-IV), the correct ones are: **[JEE Main (Sept.) 2020]**
(I) Be has smaller atomic radius compared to Mg.
(II) Be has higher ionization enthalpy than Al.
(III) Charge/radius ratio of Be is greater than that of Al.
(IV) Both Be and Al form mainly covalent compounds.
(a) (I), (III) and (IV) (b) (II), (III) and (IV)
(c) (I), (II) and (III) (d) (I), (II) and (IV)

19. The order of electron gain enthalpy in kJ/mol of fluorine, chlorine, bromine and iodine, respectively are: **[JEE Main (Jan.) 2020]**
(a) –333, – 325, – 349 and –296 (b) –333, – 349, – 325 and –296
(c) –349, – 333, – 325 and –296 (d) –296, – 325, – 333 and –349

20. Within each pair of elements F and Cl, S and Se and Li and Na, respectively, the elements that release more energy upon an electron gain are: **[JEE Main (Jan.) 2020]**
(a) F, Se and Na (b) Cl, Se and Na (c) F, S and Li (d) Cl, S and Li

21. The process that is not endothermic in nature is: **[JEE Main (Sept.) 2020]**
(a) $H(g) + e^- \longrightarrow H^-(g)$ (b) $O^-(g) + e^- \longrightarrow O^{2-}(g)$
(c) $Na(g) \longrightarrow Na^+(g) + e^-$ (d) $Ar(g) + e^- \longrightarrow Ar^-(g)$

22. The acidic, basic and amphoteric oxides, respectively, are: **[JEE Main (Jan.) 2020]**
(a) Cl_2O, CaO, P_4O_{10} (b) MgO, Cl_2O, Al_2O_3
(c) Na_2O, SO_3, Al_2O_3 (d) N_2O_3, Li_2O, Al_2O_3

23. Three elements X, Y and Z are in the 3rd period of the periodic table. The oxides of X, Y and Z, respectively, are basic, amphoteric and acidic. The correct order of the atomic numbers of X, Y and Z is: **[JEE Main (Sept.) 2020]**
(a) $X < Z < Y$ (b) $Z < Y < X$ (c) $X < Y < Z$ (d) $Y < X < Z$

24. In which of the following pairs, the outermost electronic configuration will be the same? **[JEE Main (Feb.) 2021]**
(a) Cr^+ and Mn^{2+} (b) Ni^{2+} and Cu^+ (c) Fe^{2+} and Co^+ (d) V^{2+} and Cr^+

25. The characteristics of elements X, Y and Z with atomic numbers, respectively, 33, 53 and 83 are: **[JEE Main (March) 2021]**
(a) X and Y are metalloids and Z is a metal.
(b) X is a metalloid, Y is a non-metal and Z is a metal.
(c) X, Y and Z are metals.
(d) X and Z are non-metals and Y is a metalloid.

26. The common positive oxidation states for an element with atomic number 24, are: **[JEE Main (March) 2021]**
(a) +2 to +6 (b) +1 and +3 to +6 (c) +1 and +3 (d) +1 to +6

27. The set of elements that differ in mutual relationship from those of the other sets is: **[JEE Main (March) 2021]**
(a) Li – Mg (b) B – Si (c) Be – Al (d) Li – Na

28. The ionic radius of Na^+ ions is 1.02Å. The ionic radii (in Å) of Mg^{2+} and Al^{3+} respectively, are:

[JEE Main (March) 2021]

(a) 1.05 and 0.99 (b) 0.72 and 0.54 (c) 0.85 and 0.99 (d) 0.68 and 0.72

29. The first ionization energy of magnesium is smaller as compared to that of elements X and Y, but higher than that of Z. The elements X, Y and Z, respectively, are :

[JEE Main (March) 2021]

(a) chlorine, lithium and sodium (b) argon, lithium and sodium
(c) argon, chlorine and sodium (d) neon, sodium and chlorine

30. Match List-I with List-II

	List-I (Electronic configuration of elements)		**List-II** ($\Delta_i H$ in kJ mol^{-1})
(A)	$1s^2 2s^2$	(i)	801
(B)	$1s^2 2s^2 2p^4$	(ii)	899
(C)	$1s^2 2s^2 2p^3$	(iii)	1314
(D)	$1s^2 2s^2 2p^1$	(iv)	1402

Choose the most appropriate answer from the options given below:

[JEE Main (Feb.) 2021]

(a) (A) → (ii), (B) → (iii), (C) → (iv), (D) → (i)
(b) (A) → (i), (B) → (iv), (C) → (iii), (D) → (ii)
(c) (A) → (i), (B) → (iii), (C) → (iv), (D) → (ii)
(d) (A) → (iv), (B) → (i), (C) → (ii), (D) → (iii)

31. Identify the elements X and Y using the ionisation energy values given below :

Ionization energy (kJ/mol)

	1st	2nd
X	495	4563
Y	731	1450

[JEE Main (March) 2021]

(a) X = Na; Y = Mg (b) X = Mg; Y = F (c) X = Mg; Y = Na (d) X = F; Y = Mg

32. Consider the elements Mg, Al, S, P and Si, the correct increasing order of their first ionization enthalpy is :

[JEE Main (Feb.) 2021]

(a) Mg < Al < Si < S < P (b) Al < Mg < Si < S < P
(c) Mg < Al < Si < P < S (d) Al < Mg < S < Si < P

33. The correct order of electron gain enthalpy is: **[JEE Main (Feb.) 2021]**

(a) S > Se > Te > O (b) Te > Se > S > O (c) O > S > Se > Te (d) S > O > Se > Te

34. The absolute value of the electron gain enthalpy of halogens satisfies:

[JEE Main (March) 2021]

(a) I > Br > Cl > F (b) Cl > Br > F > I (c) Cl > F > Br > I (d) F > Cl > Br > I

35. Compound *A* used as a strong oxidizing agent is amphoteric in nature. It is the part of lead storage batteries. Compound *A* is : **[JEE Main (Feb.) 2021]**

(a) PbO_2 (b) PbO (c) $PbSO_4$ (d) Pb_3O_4

36. Which pair of oxides is acidic in nature? **[JEE Main (Feb.) 2021]**

(a) B_2O_3, CaO (b) B_2O_3, SiO_2 (c) N_2O, BaO (d) CaO, SiO_2

37. The set that represents the pair of neutral oxides of nitrogen is: **[JEE Main (March) 2021]**

(a) NO and N_2O (b) N_2O and N_2O_3 (c) N_2O and NO_2 (d) NO and NO_2

38. The correct order of conductivity of ions in water is: **[JEE Main (March) 2021]**

(a) $Na^+ > K^+ > Rb^+ > Cs^+$ (b) $Cs^+ > Rb^+ > K^+ > Na^+$

(c) $K^+ > Na^+ > Cs^+ > Rb^+$ (d) $Rb^+ > Na^+ > K^+ > Li^+$

39. The IUPAC nomenclature of an element with electronic configuration [Rn] $5f^{14}6d^17s^2$ is : **[JEE Main (July) 2022 (I)]**

(a) Unnilbium (b) Unnilunium

(c) Unnilquadium (d) Unniltrium

40. The first ionization enthalpy of Na, Mg and Si, respectively, are : 496, 737 and 786 kJ mol^{-1}. The first ionization enthalpy (kJ mol^{-1}) of Al is : **[JEE Main (July) 2022 (I)]**

(a) 487 (b) 768 (c) 577 (d) 856

41. An element *A* of group 1 shows similarity to an element *B* belonging to group 2. If *A* has maximum hydration enthalpy in group 1 then *B* is : **[JEE Main (July) 2022 (II)]**

(a) Mg (b) Be (c) Ca (d) Sr

42. It is observed that characteristic X-ray spectra of elements show regularity. When frequency to the power '*n*' *i.e.*, ν^n of X-rays emitted is plotted against atomic number '*Z*', following graph is obtained. **[JEE Main (Jan.) 2023 (I)]**

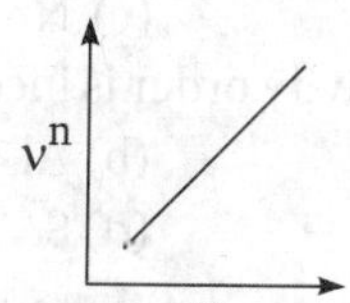

The value of '*n*' is :

(a) 1 (b) 2 (c) $\frac{1}{2}$ (d) 3

43. Which one of the following sets of ions represents a collection of isoelectronic species? (Given : Atomic number : F = 9, Cl = 17, Na = 11, Mg = 12, Al = 13, K = 19, Ca = 20, Sc = 21) **[JEE Main (Feb.) 2023 (II)]**

(a) $N^{3-}, O^{2-}, F^-, S^{2-}$ (b) $Ba^{2+}, Sr^{2+}, K^+, Ca^{2+}$

(c) $K^+, Cl^-, Ca^{2+}, Sc^{3+}$ (d) $Li^+, Na^+, Mg^{2+}, Ca^{2+}$

44. The difference between electron gain enthalpies will be maximum between : **[JEE Main (April) 2023 (I)]**

(a) Ne and F (b) Ne and Cl (c) Ar and Cl (d) Ar and F

45. The correct increasing order of the ionic radii is : **[JEE Main (Jan.) 2023 (I)]**

(a) $Cl^- < Ca^{2+} < K^+ < S^{2-}$
(b) $K^+ < S^{2-} < Ca^{2+} < Cl^-$
(c) $S^{2-} < Cl^- < Ca^{2+} < K^+$
(d) $Ca^{2+} < K^+ < Cl^- < S^{2-}$

46. The transition metal having highest 3rd ionisation enthalpy is : **[JEE Main (Jan.) 2024]**

(a) Cr (b) Mn (c) V (d) Fe

47. The correct sequence of electron gain enthalpy of the elements listed below is : **[JEE Main (Jan.) 2024]**

(A) Ar (B) Br (C) F (D) S

Choose the most appropriate from the options given below :

(a) C > B > D > A
(b) A > D > B > C
(c) A > D > C > B
(d) D > C > B > A

48. Match List-I with List-II :

List-I (Species)	List-II (Electronic distribution)
(A) Cr^{2+}	(I) $3d^8$
(B) Mn^+	(II) $3d^3 4s^1$
(C) Ni^{2+}	(III) $3d^4$
(D) V^+	(IV) $3d^5 4s^1$

Choose the correct answer from the options given below : **[JEE Main (Jan.) 2024]**

(a) A–I, B–II, C–III, D–IV
(b) A–III, B–IV, C–I, D–II
(c) A–IV, B–III, C–I, D–II
(d) A–II, B–I, C–IV, D–III

49. The element having the highest first ionization enthalpy is : **[JEE Main (Jan.) 2024]**

(a) Si (b) Al (c) N (d) C

50. Which of the following electronegativity order is incorrect? **[JEE Main (Jan.) 2025]**

(a) Al < Mg < B < N
(b) Al < Si < C < N
(c) Mg < Be < B < N
(d) S < Cl < O < F

51. Match List-I with List-II : **[JEE Main (Jan.) 2025]**

List-I		List-II	
(A)	$Al^{3+} < Mg^{2+} < Na^+ < F^-$	(I)	Ionisation Enthalpy
(B)	B < C < O < N	(II)	Metallic character
(C)	B < Al < Mg < K	(III)	Electronegativity
(D)	Si < P < S < Cl	(IV)	Ionic radii

Choose the correct answer from the options given below :

(a) A-IV, B-I, C-III, D-II
(b) A-II, B-III, C-IV, D-I
(c) A-IV, B-I, C-II, D-III
(d) A-III, B-IV, C-II, D-I

52. The element that does not belong to the same period of the remaining elements (modern periodic table) is : **[JEE Main (Jan.) 2025]**

(a) Palladium (b) Iridium (c) Osmium (d) Platinum

53. Which of the following statements are NOT true about the periodic table?

A. The properties of elements are function of atomic weights.
B. The properties of elements are function of atomic weights.
C. Elements having similar outer electronic configuration are arranged in same period.
D. An element's location reflects the quantum numbers of the last filled orbital.
E. The number of elements in a period is same as the number of atomic orbitals available in energy level that is being filled.

Choose the correct answer from the options given below : **[JEE Main (Jan.) 2025]**

(a) A, C and E only (b) D and E only
(c) A and E only (d) B, C and E only

54. Match **List-I** with **List-II.** **[JEE Main (Jan.) 2025]**

	List-I (Transition metal ion)		**List-II [Spin only magnetic moment (B.M.)]**
(A)	Ti^{3+}	(I)	3.87
(B)	V^{2+}	(II)	0.00
(C)	Ni^{2+}	(III)	1.73
(D)	Sc^{3+}	(IV)	2.84

Choose the correct answer from the options given below :

(a) (A)-(III), (B)-(I), (C)-(II), (D)-(IV) (b) (A)-(III), (B)-(I), (C)-(IV), (D)-(II)
(c) (A)-(IV), (B)-(II), (C)-(III), (D)-(I) (d) (A)-(II), (B)-(IV), (C)-(I), (D)-(III)

55. The successive 5 ionisation energies of an element are 800, 2427, 3658, 25024 and 32824 kJ/mol, respectively. By using the above values predict the group in which the above element is present : **[JEE Main (Jan.) 2025]**

(a) Group 2 (b) Group 13 (c) Group 4 (d) Group 14

56. The incorrect decreasing order of atomic radii is: **[JEE Main (Jan.) 2025]**

(a) Mg > Al > C > O (b) Al > B > N > F (c) Be > Mg > Al > Si (d) Si > P > Cl > F

57. An element 'E' has the ionisation enthalpy value of 374 kJ mol^{-1}. 'E' reacts with elements A, B, C and D with electron gain enthalpy value of –328, – 349, – 325 and –295 kJ mol^{-1} respectively. The correct order of the products E_A, E_B, E_C and E_D in terms of ionic character is: **[JEE Main (Jan.) 2025]**

(a) $E_B > E_A > E_C > E_D$ (b) $E_D > E_C > E_A > E_B$
(c) $E_A > E_B > E_C > E_D$ (d) $E_D > E_C > E_B > E_A$

58. First ionisation enthalpy values of first four group 15 elements are given below. Choose the correct value for the element that is a main component of apatite family: **[JEE Main (Jan.) 2025]**

(a) 1012 kJ mol^{-1} (b) 1402 kJ mol^{-1}
(c) 834 kJ mol^{-1} (d) 947 kJ mol^{-1}

B. Statement and Explanation Type Problems

Read the following questions and answer as per the direction given below :

(a) Statement 1 is true; Statement 2 is true; Statement 2 is the correct explanation of Statement 1.

(b) Statement 1 is true; Statement 2 is true; Statement 2 is not the correct explanation of Statement 1.

(c) Statement 1 is true; Statement 2 is false.

(d) Statement 1 is false; Statement 2 is true.

(e) Both Statement 1 and Statement 2 are false.

1. **Statement 1 :** F atom has a less negative electron affinity than Cl atom.
   Statement 2: Additional electrons are repelled more effectively by 3 *p* electrons in Cl atom than by 2 *p* electrons in F atom. **(IIT 1998)**
2. **Statement 1 :** LiCl is predominantly a covalent compound.
   Statement 2 : Electronegativity difference between Li and Cl is too small. **(IIT 1998)**
3. **Statement 1 :** The first ionisation energy of Be is greater than that of B.
   Statement 2 : 2 *p* orbital is lower in energy than 2 *s*. **(IIT 2000)**
4. **Statement 1 :** An element in the extreme left of the periodic table forms acidic oxides.
   Statement 2 : Acid is formed during the reaction between water and oxide of a reactive element present in the extreme right of the periodic table. **[JEE Main (Jan.) 2025]**
5. **Statement 1 :** A plot of $\sqrt{\nu}$ (ν = frequency of X-rays emitted) :
   Statement 2 : A plot of ν (ν = frequency of X-rays emitted) *vs* atomic number is a straight line.
   [JEE Main (Jan.) 2025]
6. **Statement 1 :** According to the Law of Octaves, the elements were arranged in the increasing order of their atomic number.
   Statement 2 : Meyer observed a periodically repeated pattern upon plotting physical properties of certain elements against their respective atomic numbers. **[JEE Main (Jan.) 2025]**
7. **Statement 1 :** The radii of isoelectronic species increases in the order.
   $$Mg^{2+} < Na^{+} < F^{-} < O^{2-}$$
   Statement 2 : The magnitude of electron gain enthalpy of halogen decreases in the order.
   $$Cl > F > Br > I$$
   [JEE Main (Jan.) 2025]

C. Integer Answer Type Problems

1. The periodic table consists of 18 groups. An isotope of copper, on bombardment with protons, undergoes a nuclear reaction yielding element *X* as shown below. To which group, element *X* belongs in the periodic table ? **(IIT 2012)**
   $$^{63}_{29}Cu + ^{1}_{1}H \longrightarrow 6\,^{1}_{0}n + \alpha + 2\,^{1}_{1}H + X$$
2. The atomic number of Unnilunium is ____________. **[JEE Main (Sept.) 2020]**
3. Number of amphoteric compound among the following is _______ **[JEE Main (Feb.) 2021]**
   (a) BeO (b) BaO (c) $Be(OH)_2$ (d) $Sr(OH)_2$

4. The total number of isoelectronic species from the given set is ________.
$O^{2-}, F^-, Al, Mg^{2+}, Na^+, O^+, Mg, Al^{3+}, F$ **[JEE Main (April) 2023 (I)]**

5. Total number of acidic oxides among
$N_2O_3, NO_2, N_2O, Cl_2O_7, SO_2$, CO, CaO, Na_2O and NO is ________. **[JEE Main (Jan.) 2023 (II)]**

6. The 'Spin only' Magnetic moment for $[Ni(NH_3)_6]^{2+}$ is ______ $\times 10^{-1}$ BM.
(Given = Atomic number of Ni = 28) **[JEE Main (Jan.) 2024]**

7. If IUPAC name of an element is "Unununium" then the element belongs to *n*th group of periodic table. The value of *n* is ________. **[JEE Main (Jan.) 2024]**

Answers

[A] Objective Questions (Only one option is correct)

1. (b)	**2.** (a)	**3.** (d)	**4.** (c)	**5.** (d)	**6.** (b)	**7.** (b)	**8.** (c)	**9.** (c)	**10.** (d)
11. (b)	**12.** (d)	**13.** (d)	**14.** (b)	**15.** (d)	**16.** (d)	**17.** (c)	**18.** (d)	**19.** (b)	**20.** (d)
21. (a)	**22.** (d)	**23.** (c)	**24.** (a)	**25.** (b)	**26.** (a)	**27.** (d)	**28.** (b)	**29.** (c)	**30.** (a)
31. (a)	**32.** (b)	**33.** (a)	**34.** (c)	**35.** (a)	**36.** (b)	**37.** (a)	**38.** (b)	**39.** (d)	**40.** (c)
41. (a)	**42.** (c)	**43.** (c)	**44.** (b)	**45.** (d)	**46.** (b)	**47.** (b)	**48.** (b)	**49.** (c)	**50.** (a)
51. (c)	**52.** (a)	**53.** (a)	**54.** (b)	**55.** (b)	**56.** (c)	**57.** (a)	**58.** (a)		

[B] Statement and Explanation Type Problems

1. (c) **2.** (c) **3.** (c) **4.** (d) **5.** (e) **6.** (e) **7.** (a)

[C] Integer Type Problems

1. (8)	**2.** (101)	**3.** (2)	**4.** (5)	**5.** (4)
6. (28)	**7.** (11)			

Hints and Solutions

A. Objective Questions

[Only one option is correct]

1. (b) E.A. = Ionisation potential

$\therefore$ E.A. of $Na^+ = -5.1\,eV$

2. (a) The ionic radii of isoelectronic ions increase with the decrease in magnitude of the nuclear charge.

$$\underset{1.36\text{Å}}{F^-} < \underset{1.40\text{Å}}{O^{2-}} < \underset{1.71\text{Å}}{N^{3-}}$$

3. (d) Rb, Na and K belong to group 1 which on losing one electron attain noble gas configuration and thus, have low value of ionisation energies. On the other hand, Sc with electronic configuration $[Ar]3d^1 4s^2$ loses its electron from 4*s*-orbital thus, it does not achieve that much stable configuration as in group 1 elements. Hence, it shows higher first ionisation energy.

4. (c) Due to diagonal relationship, both Li and Mg display some similar properties, but in the case of carbonates, Mg can form basic carbonates such as $3MgCO_3 \cdot Mg(OH)_2 \cdot 3H_2O$. In contrast, Li only form typical carbonate Li_2CO_3 as other alkali metals. It does not form any basic carbonate having both carbonate and hydroxide ions.

5. (d) The IUPAC symbol for element with atomic number 119 would be uue.

6. (b) Element with atomic number $(Z) = 101$ is Mendelevium (Md). It belongs to actinide series. While element with atomic number $(Z) = 104$ is Rutherfordium (Rf) it belongs to group 4 of periodic table.

7. (b) $Z = 21$ Scandium
$Z = 25$ Manganese
$Z = 42$ Molybdenum
$Z = 72$ Hafnium
All are transition elements.

8. (c) Atomic radius of Ag is closest to Au. Silver belongs to 5th period while gold belongs to 6th period. As we move from Ag to Au, $4f$ electrons are added which have poor shielding power. So, due to lanthanoid contraction size remain almost constant of that of Ag and Au.

9. (c)

Atomic radius decreases →

C   O   F
           Cl   Atomic radius increases ↓
           Br

Order will be: Br > Cl > C > O > F

10. (d) Given species are isoelectronic species, all have 10 electrons.
For isoelectronic speices as z increases, ionic radii decreases.
So order will be : $O^{2-} > F^- > Na^+ > Mg^{2+}$

11. (b) For isoelectronic species size decreases on increasing atomic number (z). Order of ionic radii will be :

$$N^{3-} > O^{2-} > F^- > Na^+ > Mg^{2+} > Al^{3+}$$

12. (d)

Element	Electronic configuration
Ni	$[Ar]\,4s^2 3d^8$
Co	$[Ar]\,4s^2 3d^7$
Mn	$[Ar]\,4s^2 3d^5$
Fe	$[Ar]\,4s^2 3d^6$

Fe have minimum 3rd ionization energy, since after removing three electrons it attain stable configuration $[Ar]\,3d^5$.
(Half-filled orbitals have extra stability).

13. (d)

Element	Electronic configuration
Na	$[Ne]\,3s^1$
Mg	$[Ne]\,3s^2$
Al	$[Ne]\,3s^2 3p^1$
Si	$[Ne]\,3s^2 3p^2$

Moving in a period I.E. increases, Mg have greater I.E. than Al : Na < Al < Mg < Si

14. (b) Electronic configuration :

$$_4Be \longrightarrow 1s^2\,2s^2$$

$$_5Be \longrightarrow 1s^2\,2s^2\,2p^1$$

Statement (i), (ii) and (iii) are correct.
Atomic radius of Be is more than B.

15. (d) Since second ionisation energy is very high it means metal have single electron in its valence shell. After removing its single electron it attain stable noble gas configuration. So formula of its hydroxide will be MOH.

$$\underset{(1\text{ mol})}{MOH} + \underset{(1\text{ mol})}{HCl} \longrightarrow MCl + H_2O$$

$$\underset{\left(\frac{1}{2}\text{ mol}\right)}{MOH} + \underset{(1\text{ mol})}{\frac{1}{2}H_2SO_4} \longrightarrow \frac{1}{2}M_2SO_4 + H_2O$$

16. (d) Since ionization enthalpy increases abruptly after third ionisation enthalpy it means it has 3 valence electrons. After removal of 3 valence electrons it attains stable noble gas configuration.

17. (c) In a period moving from left to right ionization enthalpy, electron gain enthalpy, electronegativity increases white atomic radius decrease due to increased nuclear charge.

18. (d) ❖ Moving down the group size (Atomic radius) increases. So, radius (Be) < radius (Mg).
❖ Be have smaller size than Al, so it has higher ionisation energy.
❖ Charge/ratio of Be ≅ Al. Since both are related with diagonal relationship.
❖ Both boron and aluminium have high sum of total of first three ionisation enthalpy $(IE_1 + IE_2 + IE_3)$. So, they cannot lose electrons easily to form B^{3+} and Al^{3+} ions. That is reason both Be and Al form mainly covalent bonds.

19. (b) Since correct order of electron gain enthalpy is : $Cl > F > Br > I$

20. (d) As we move down the group, electron affinity decreases since size increases

So, E. A. (S) > E. A. (Se)

E. A. (Li) > E. A. (Na)

Also, E. A. (Cl) > E. A. (F)

21. (a) $\underset{(1s^1)}{H(g)} + e^- \longrightarrow \underset{(1s^2)}{H^-(g)}$

Due to formation of duplet (stable configuration) process is exothermic. In option (b), $O^-(g)$ experiences repulsive force from incoming electron so process is endothermic.

In option (c), electron is removed from Na, so energy is required (endothermic).

In option (d), Ar has stable noble gas configuration, So adding electron to it will be endothermic process.

22. (d) $N_2O_3 \longrightarrow$ Acidic oxide

$Li_2O \longrightarrow$ Basic oxide

$Al_2O_3 \longrightarrow$ Amphoteric acid

23. (c) In moving a period from left to right oxides character change from basic $\longrightarrow$ amphoteric $\longrightarrow$ acidic

So, order of atomic number will be $X < Y < Z$.

24. (a)

Ions	Electronic configuration	Ions	Electronic configuration
Mn^{2+}	$[Ar]3d^5$	Fe^{2+}	$[Ar]3d^6$
Cr^+	$[Ar]3d^5$	Co^+	$[Ar]3d^7 4s^1$
Ni^{2+}	$[Ar]3d^8$	V^{2+}	$[Ar]3d^3$
Cu^+	$[Ar]3d^{10}$		

25. (b)

Atomic number	Element	Nature
33	Arsenic (As)	Metalloid
53	Iodine (I)	Non-metal
83	Bismuth (Bi)	Metal

26. (a) Element with atomic number $(Z) = 24$ is chromium (Cr). It shows oxidation state from +2 to +6 .

27. (d) Elements related through diagonal relationship are Li — Mg, B — Si, Be — Al. But Li & Na don't show diagonal relationship.

28. (b) For isoelectronic species, ionic radii decreases on increasing atomic number (Z). So, order will be:

$$Na^{+} > Mg^{2+} > Al^{3+}$$

29. (c) Order of 1st I.E. of 3rd period elements :

$$Na < Al < Mg < Si < S < P < Cl < Ar$$

So, $X = Ar, Y = Cl, Z = Na$.

30. (a)

Element	Electronic configuration
Be	$1s^2 2s^2$
O	$1s^2 2s^2 2p^4$
N	$1s^2 2s^2 2p^3$
B	$1s^2 2s^2 2p^1$

Order of ionisation enthalpy $B < Be < O < N$

As we move in a period ionisation energy increases. But Be and N have exceptional value of ionisation enthalpy due to extra stability of completely filled and half filled orbitals.

31. (a) In element 'X' IE_1 is low while IE_2 is very high it means it leave single electron in its valence shell and after removing it, attains noble gas configuration, so it should be sodium (Na). While element B has IE_1 and IE_2 low, it means it has 2 valence electrons so element Y should be Mg.

32. (b)

Element	Electronic configuration
Mg	$[Ne]3s^2$
Al	$[Ne]3s^2 3p^1$
Si	$[Ne]3s^2 3p^2$
P	$[Ne]3s^2 3p^3$
S	$[Ne]3s^2 3p^4$

As we proceed from left from right ionisation enthalpy increases, Mg and P have exceptional behaviour due to extra stability of half filled and fully filled orbitals.

Correct sequence is :

$$Al < Mg < Si < S < P$$

33. (a) In a group moving from top to down electron gain enthalpy decreases since size increases. But oxygen show least electron gain enthalpy due to its small size and high charge density. So correct order is : $S > Se > Te > O$

34. (c) As we go down in a group electron gain enthalpy of atom decreases since added electron feels lesser electrostatic force of attraction to nucleus due to larger atomic size.

Fluorine show lesser electron gain enthalpy than chlorine since incoming electron experiences more repulsive force in fluorine due to its small size.

So, correct order of electron gain enthalpy is :

$$Cl > F > Br > I$$

35. (a) PbO_2 is amphoteric in nature and it is strong oxidizing agent since lead (Pb) is in highest oxidation state (+4).
It is also part of lead storage battery.

36. (b) Oxides of non-metal are acidic in nature.

37. (a)

N_2O	Neutral oxide
NO	Neutral oxide
NO_2	Acidic oxide
N_2O_3	Acidic oxide

38. (b) As we move down in a group, extent of hydration of ions decreases and conductivity increases. So, correct order is :

$$Cs^+ > Rb^+ > K^+ > Na^+$$

39. (d) Atomic no. of the element [Rn] $5f^{14}6d^17s^2 = 86 + 14 + 1 + 2 = 103$

IUPAC name = Unniltrium

40. (c) I.E. order : Na < Al < Mg < Si

$\therefore \quad 496 < \text{I.E. (Al)} < 737 < 786$

Only option (c), matches the condition.

41. (a) Li Mg Diagonal relationship

Na Mg (Li ⟶ Mg diagonal arrow)

$Li^+ \longrightarrow$ Maximum hydration enthalpy in group 1 due to small size. So '*B*' is Mg.

42. (c) According to Henry Moseley $\sqrt{v} \propto Z - b$

So, $\quad n = \dfrac{1}{2}$

43. (c) $K^+(18)$, $Cl^-(18)$, $Ca^{2+}(18)$, $Sc^{3+}(18)$

44. (b) Cl has the most negative ΔH_{eg} among all the elements and Ne has the most positive ΔH_{eg}.

45. (d) In isoelectronic species size $\propto \dfrac{1}{Z}$. Hence correct increasing order of ionic radii is :

$$Ca^{2+} < K^+ < Cl^- < S^{2-}$$

46. (b) 3rd Ionisation energy : [NCERT Data]

V : 2833 kJ/mol Cr : 2990 kJ/mol

Mn : 3260 kJ/mol Fe : 2962 kJ/mol

Alternative

Mn : $3d^5 4s^2$ Fe : $3d^6 4s^2$

Cr : $3d^5 4s^1$ V : $3d^3 4s^2$

So, Mn has highest 3rd I.E. among all the given elements due to d^5 configuration.

47. (b)

Element	$\Delta_{eg}H$ **(kJ/mol)**
F	–333
S	–200
Br	–325
Ar	+96

48. (b) $_{24}Cr \longrightarrow [Ar]\,3d^5 4s^1$; $Cr^{2+} \longrightarrow [Ar]\,3d^4$

$_{25}Mn \longrightarrow [Ar]\,3d^5 4s^2$; $Mn^{+} \longrightarrow [Ar]\,3d^5 4s^1$

$_{28}Ni \longrightarrow [Ar]\,3d^8 4s^2$; $Ni^{2+} \longrightarrow [Ar]\,3d^8$

$_{23}V \longrightarrow [Ar]\,3d^3 4s^2$; $V^{+} \longrightarrow [Ar]\,3d^3 4s^1$

49. (c) Al < Si < C < N; IE_1 order.

50. (a)

	Li	Be	B	C	N	O	F
(E.N.) =	1	1.5	2	2.5	3	3.5	4.0

On pauling scale

	Na	Mg	Al	Si	P	S	Cl
(E.N.) =	0.9	1.2	1.5	1.8	2.1	2.5	3.0

Correct order Mg < Al < B < N

51. (c) Ionic radii : $Al^{3+} < Mg^{2+} < Na^{+} < F^{-}$

Ionisation energy : B < C < O < N

Metallic character : B < Al < Mg < K

Electron negativity : Si < P < S < Cl

52. (a) Palladium ⇒ 5th period

Iridium, Osmium, Platinum ⇒ 6th period

53. (a) Properties of elements are periodic function of their atomic number. Elements having similar outer electronic configuration are arranged in same group. Number of elements in a period is not equal to number of atomic orbitals available in energy level that is being filled.

Hence, A, C and E are incorrect.

54. (b) $Sc^{3+} = 3d^0$ $\quad \therefore \mu_{spin} = 0$

$V^{2+} = 3d^3$ $\quad \therefore \mu_{spin} = 3.87$ B.M.

$Ni^{2+} = 3d^8$ $\quad \therefore \mu_{spin} = 2.84$ B.M.

$Ti^{3+} = 3d^1$ $\quad \therefore \mu_{spin} = 1.73$ B.M.

55. (b) The IE_4 is suddenly very high therefore element must have 3 valence $e^-(s)$ and it belong to group 13.

56. (c) Correct order of atomic radii : Be < Mg > Al > Si

57. (a) Difference between I.E. and E.G.E. increases, ionic character increases.

B. Statement and Explanation Type Problems

1. (c) Statement-1 is a fact while statement-2 is false. Actually additional electrons are more effectively repelled by $2p$ in F atom than by $3p$ in Cl.

2. (c) Covalent character in LiCl is incorporated due to polarisation of anion by cation (Fajan's rule)

3. (c) Statement-1 is a fact while statement-2 is false. For subshells belonging to same shell, the order of energy is $s < p < d < f$, thus energy of $2s < 2p$.

4. (d) On moving left to right in periodic table non-metallic character increases and we know that non-metal oxides are acidic in nature.

Non-metallic character ↑ Acidic strength of oxide ↑

5. (e)

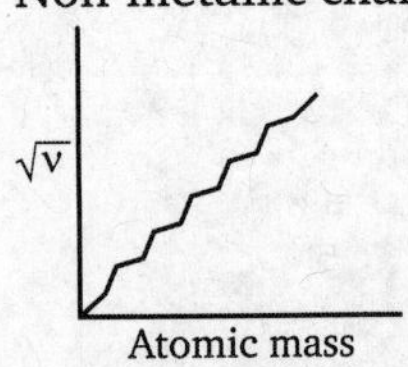

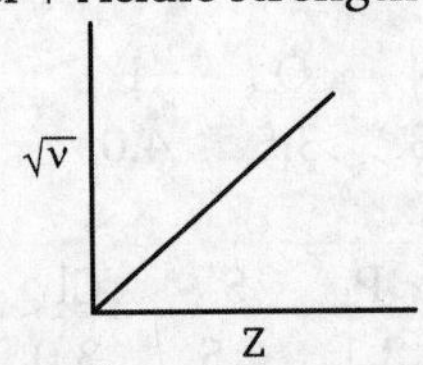

6. (e) Law of octaves was arranged in the increasing order of their atomic weight.

Lothar Meyer plotted the physical properties such as atomic volume, melting point and boiling point against atomic weight.

7. (a) (i) For isoelectronic species –ve charge increases, radii increases.

(ii) Magnitude of E.G.E. : Cl > F > Br > I

C. Integer Type Problems

1. (8) $^{63}_{29}\text{Cu} + {}^{1}_{1}\text{H} \longrightarrow 6\,{}^{1}_{0}n + {}^{4}_{2}\text{He}(\alpha) + 2\,{}^{1}_{1}\text{H} + {}^{52}_{26}X$

Atomic number 26 represents Fe which belongs to group 8.

Alternative path : $Z = 26$

Electronic configuration $= 1s^2 2s^2 2p^6 3s^2 3p^6 4s^2 3d^6$

Thus element X belongs to d-block and for d-block,

Group number = electrons in $(n-1)$ subshell number of electrons in valence shell

$= 6 + 2 = 8$

2. (101)

Un ⟶ 1

nil ⟶ 0

Unnilunium ⟶ 101

3. (2)

Compound	**Nature**
BeO	Amphoteric
BaO	Basic
$Be(OH)_2$	Amphoteric
$Sr(OH)_2$	Basic

4. (5) Isoelectronic species $O^{2-}, F^{-}, Mg^{2+}, Na^{+}, Al^{3+}$ (all have $10\ e^{-}$)

5. (4) Acidic oxides are $N_2O_3, NO_2, Cl_2O_7, SO_2$
Basic oxides are CaO, Na_2O
Neutral oxides are CO, NO, N_2O

6. (28) NH_3 act as WFL with Ni^{2+}

$Ni^{2+} = 3d^8$

⇅	⇅	⇅	↑	↑

Number of unpaired electron = 2

$$\mu = \sqrt{n(n+2)} = \sqrt{8} = 2.82 \text{ BM}$$
$$= 28.2 \times 10^{-1} \text{BM}$$
$$x = 28$$

7. (11) 111 belongs to 11th group.

❑❑❑

CHAPTER 2

Chemical Bonding and Molecular Structure

A. Objective Questions

[Only one option is correct]

1. The bond energy (in kcal mol^{-1}) of C — C single bond is approximately : **(IIT 2010)**

(a) 1 (b) 10 (c) 100 (d) 1000

2. The species having pyramidal shape is : **(IIT 2010)**

(a) SO_3 (b) BrF_3 (c) SiO_3^{2-} (d) OSF_2

3. Which one of the following molecules is expected to exhibit diamagnetic behaviour? **[JEE (Mains) 2013]**

(a) C_2 (b) N_2 (c) O_2 (d) S_2

4. In which of the following pairs of molecules/ions, both the species are not likely to exist? **[JEE (Mains) 2013]**

(a) H_2^+, He_2^{2-} (b) H_2^-, He_2^{2-} (c) H_2^{2+}, He_2 (d) H_2^-, He_2^{2+}

5. Stability of the species Li_2, Li_2^- and Li_2^+ increases in the order of : **[JEE (Mains) 2013]**

(a) $Li_2 < Li_2^+ < Li_2^-$ (b) $Li_2^- < Li_2^+ < Li_2$

(c) $Li_2 < Li_2^- < Li_2^+$ (d) $Li_2^- < Li_2 < Li_2^+$

6. Which one of the following properties is **not** shown by NO ? **[JEE (Mains) 2014]**

(a) It is diamagnetic in gaseous state

(b) It is a neutral oxide

(c) It combines with oxygen to form nitrogen dioxide

(d) It's bond order is 2.5

7. For which of the following molecule significant $\mu \neq 0$? **[JEE (Mains) 2014]**

(1) 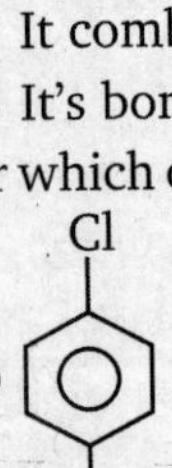
(2) 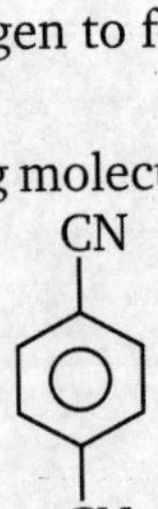
(3) 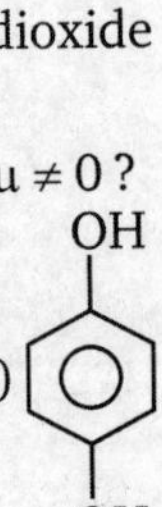
(4) 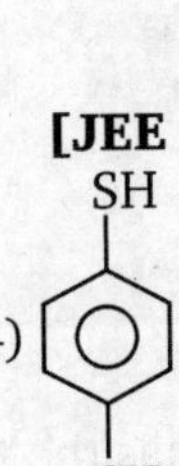

(a) Only (1) (b) (1) and (2) (c) Only (3) (d) (3) and (4)

8. Assuming $2s-2p$ mixing is not operative, the paramagnetic species among the following is : **[JEE (Advanced) 2014]**

(a) Be_2 (b) B_2 (c) C_2 (d) N_2

9. The intermolecular interaction that is dependent on the inverse cube of distance between the molecules is : **[JEE (Mains) 2015]**

(a) London force (b) hydrogen bond
(c) ion-ion interaction (d) ion-dipole interaction

10. Which of the following species is not paramagnetic? **[JEE (Mains) 2017]**

(a) O_2 (b) B_2 (c) NO (d) CO

11. Total number of lone pair of electrons in I_3^- ion is : **[JEE (Mains) 2018]**

(a) 9 (b) 12 (c) 3 (d) 6

12. Which of the following compounds contain(s) no covalent bond(s)?

$KCl, PH_3, O_2, B_2H_6, H_2SO_4$ **[JEE (Mains) 2018]**

(a) KCl (b) KCl, B_2H_6
(c) KCl, B_2H_6, PH_3 (d) KCl, H_2SO_4

13. According to molecular orbital theory which of the following will not be a viable molecule? **[JEE (Mains) 2018]**

(a) H_2^- (b) H_2^{2-} (c) He_2^{2+} (d) He_2^+

14. The correct statement about ICl_5 and ICl_4^- is : **[JEE (Mains) 2019]**

(a) both are isostructural
(b) ICl_5 is square pyramidal and ICl_4^- is square planar.
(c) ICl_5 is trigonal bipyramidal and ICl_4^- is tetrahedral.
(d) ICl_5 is square pyramidal and ICl_4^- is tetrahedral.

15. Among the following molecules/ions,

$$C_2^{2-}, N_2^{2-}, O_2^{2-}, O_2$$

which one is diamagnetic and has the shortest bond length? **[JEE (Mains) 2019]**

(a) C_2^{2-} (b) O_2 (c) N_2^{2-} (d) O_2^{2-}

16. The intermolecular potential energy for the molecules *A*, *B*, *C* and *D* given below suggest that: **[JEE Main (Sept.) 2020]**

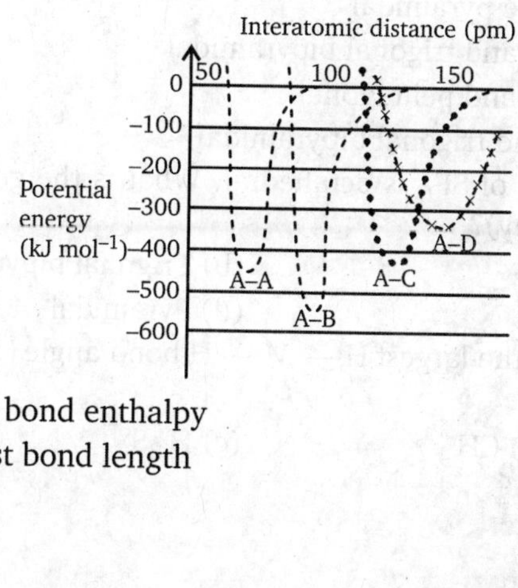

(a) *A—A* has the largest bond enthalpy
(b) *A—D* has the shortest bond length

(c) D is more electronegative than other atoms
(d) A—B has the stiffest bond.

17. The potential energy curve for the H_2 molecule as a function of internuclear distance is : **[JEE Main (Sept.) 2020]**

(a) 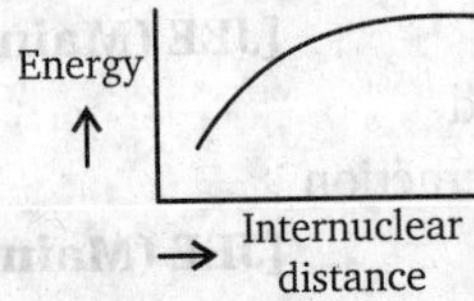

(b) 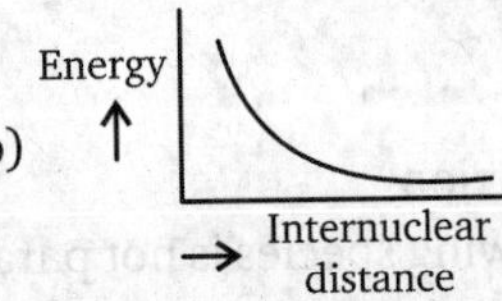

(c) 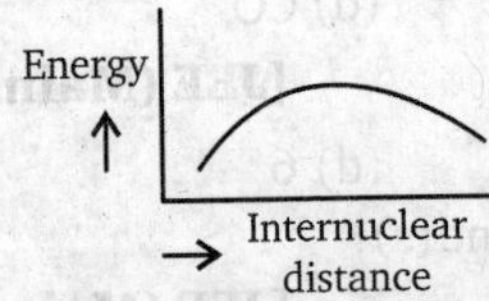

(d) 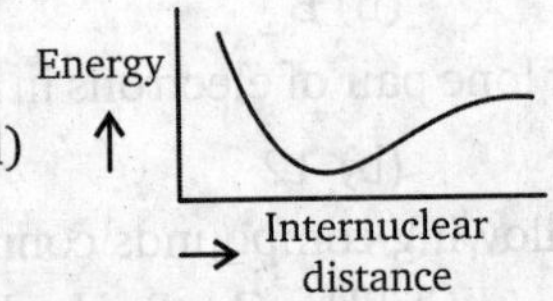

18. Arrange the following bonds according to their average bond energies in descending order: **[JEE Main (Jan.) 2020]**

C — Cl, C — Br, C — F, C — I

(a) C — F > C — Cl > C — Br > C — I (b) C — I > C — Br > C — Cl > C — F
(c) C — Cl > C — Br > C — I > C — F (d) C — Br > C — I > C — Cl > C — F

19. The molecule in which hybrid MOs involve only one d-orbital of the central atom is : **[JEE Main (Sept.) 2020]**

(a) XeF_4 (b) $[CrF_6]^{3-}$ (c) $[Ni(CN)_4]^{2-}$ (d) BrF_5

20. The correct statement with respect to dinitrogen is: **[JEE Main (Sept.) 2020]**

(a) N_2 is paramagnetic in nature
(b) it can combine with dioxygen at 25°C
(c) liquid dinitrogen is not used in cryosurgery
(d) it can be used as an inert diluent for reactive chemicals

21. If AB_4 molecule is a polar molecule, a possible geometry of AB_4 is: **[JEE Main (Sept.) 2020]**

(a) rectangular planar (b) square pyramidal
(c) tetrahedral (d) square planar

22. The shape/structure of $[XeF_5]^-$ and XeO_3F_2, respectively are : **[JEE Main (Sept.) 2020]**

(a) octahedral and square pyramidal
(b) trigonal bipyramidal and trigonal bipyramidal
(c) trigonal bipyramidal and pentagonal
(d) pentagonal planar and trigonal bipyramidal

23. The molecular geometry of SF_6 is octahedral. What is the geometry of SF_4 (including lone pair(s) of electrons, if (any)? **[JEE Main (Sept.) 2020]**

(a) Tetrahedral (b) Trigonal bipyramidal
(c) Square planar (d) Pyramidal

24. The compound that has the largest H — M — H bond angle (M = N, O, S, C), is: **[JEE Main (Sept.) 2020]**

(a) H_2O (b) CH_4 (c) H_2S (d) NH_3

25. Hydrogen peroxide, in the pure state, is: **[JEE Main (Sept.) 2020]**
(a) non-planar and colourless (b) planar and blue in colour
(c) linear and blue in colour (d) linear and almost colourless

26. The dipole moments of CCl_4, $CHCl_3$ and CH_4 are in the order: **[JEE Main (Jan.) 2020]**
(a) $CHCl_3 < CH_4 = CCl_4$ (b) $CCl_4 < CH_4 < CHCl_3$
(c) $CH_4 = CCl_4 < CHCl_3$ (d) $CH_4 < CCl_4 < CHCl_3$

27. The relative strength of interionic/intermolecular forces in decreasing order is: **[JEE Main (Jan.) 2020]**
(a) ion-dipole > dipole-dipole > ion-ion (b) dipole-dipole > ion-dipole > ion-ion
(c) ion-ion > ion-dipole > dipole-dipole (d) ion-dipole > ion-ion > dipole-dipole

28. A graph of vapour pressure and temperature for three different liquids *X*, *Y* and *Z* is shown below: **[JEE Main (Jan.) 2020]**

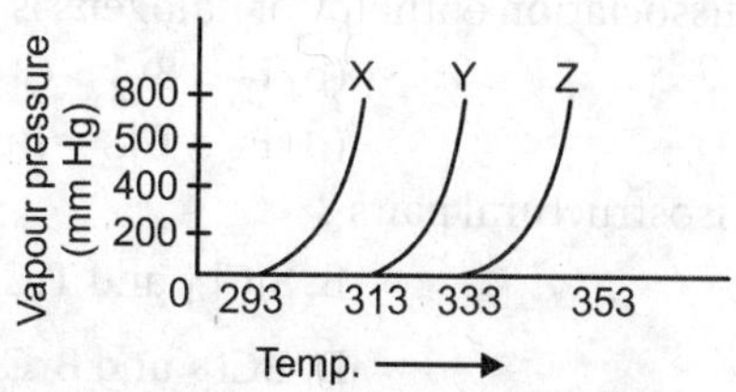

The following inferences are made :
(A) *X* has higher intermolecular interactions compared to *Y*.
(B) *X* has lower intermolecular interactions compared to *Y*.
(C) *Z* has lower intermolecular interactions compared to *Y*.
The correct inference(s) is/are:
(a) (C) (b) (A) and (C) (c) (B) (d) (A)

29. If the boiling point of H_2O is 373 K, the boiling point of H_2S will be : **[JEE Main (Sept.) 2020]**
(a) more than 373 K
(b) less than 300 K
(c) greater than 300 K but less than 373 K
(d) equal to 373 K

30. The structure of PCl_5 in the solid state is: **[JEE Main (Sept.) 2020]**
(a) square planar $[PCl_4]^+$ and octahedral $[PCl_6]^-$
(b) trigonal bipyramidal
(c) tetrahedral $[PCl_4]^+$ and octahedral $[PCl_6]^-$
(d) square pyramidal

31. Among the sulphates of alkaline earth metals, the solubilities of $BeSO_4$ and $MgSO_4$ in water, respectively, are: **[JEE Main (Sept.) 2020]**
(a) poor and poor (b) high and poor
(c) high and high (d) poor and high

32. Of the species NO, NO^+, NO^{2+} and NO^-, the one with minimum bond strength is: **[JEE Main (Sept.) 2020]**

(a) NO^+ (b) NO (c) NO^{2+} (d) NO^-

33. Oxidation number of potassium in K_2O, K_2O_2 and KO_2 respectively is: **[JEE Main (Jan.) 2020]**

(a) +1, +2 and +4 (b) +2, +1 and +½
(c) +1, + 1 and +1 (d) +1, + 4 and + 2

34. The bond order and the magnetic characteristics of CN^- are: **[JEE Main (Jan.) 2020]**

(a) 3, paramagnetic (b) 2½, diamagnetic
(c) 3, diamagnetic (d) 2½, paramagnetic

35. If the magnetic moment of a dioxygen species is 1.73 B.M, it may be: **[JEE Main (Jan.) 2020]**

(a) O_2, O_2^- or O_2^+ (b) O_2 or O_2^- (c) O_2 or O_2^+ (d) O_2^- or O_2^+

36. The correct order of bond dissociation enthalpy of halogens is : **[JEE Main (Feb.) 2021]**

(a) $Cl_2 > F_2 > Br_2 > I_2$ (b) $I_2 > Br_2 > Cl_2 > F_2$
(c) $Cl_2 > Br_2 > F_2 > I_2$ (d) $F_2 > Cl_2 > Br_2 > I_2$

37. Which of the following are isostructural pairs ? **[JEE Main (Feb.) 2021]**

A. SO_4^{2-} and CrO_4^{2-} B. $SiCl_4$ and $TiCl_4$
C. NH_3 and NO_3^- D. BCl_3 and $BrCl_3$

(a) C and D only (b) A and B only (c) A and C only (d) B and C only

38. The correct shape and I — I — I bond angles respectively in I_3^- ion are: **[JEE Main (Feb.) 2021]**

(a) Distorted trigonal planar; 135° and 90° (b) T-shaped; 180° and 90°
(c) Trigonal planar; 120° (d) Linear; 180°

39. Which among the following species has unequal bond lengths ? **[JEE Main (Feb.) 2021]**

(a) BF_4^- (b) XeF_4 (c) SF_4 (d) SiF_4

40. Given below are two statements: one is labelled as Assertion A and the other is labelled as Reason R. **[JEE Main (Feb.) 2021]**

Assertion A : In TlI_3, isomorphous to CsI_3, the metal is present in +1 oxidation state.

Reason R : Tl metal has fourteen *f* electrons in the electronic configuration.

In the light of the above statements, choose the most appropriate answer from the options given below:

(a) A is correct but R is not correct
(b) Both A and R are correct and R is the correct explanation of A.
(c) A is not correct but R is correct
(d) Both A and R are correct but R is not the correct explanation of A.

41. Given below are two statements : one is labelled as Assertion A and the other is labelled as Reason R :

Assertion A : The H — O — H bond angle in water molecule is 104.5°.

Reason R : The lone pair-lone pair repulsion of electrons is higher than the bond pair-bond pair repulsion. **[JEE Main (March) 2021]**

(a) A is false but R is true
(b) Both A and R are true, but R is not the correct correct explanation of A

(c) A is true but R is false

(d) Both A and R are true, and R is the correct explanation of A

42. A central atom in a molecule has two lone pairs of electrons and forms three single bonds. The shape of this molecule is: **[JEE Main (March) 2021]**

(a) see-saw (b) planar triangular (c) T-shaped (d) trigonal pyramidal

43. The correct statement about B_2H_6 is: **[JEE Main (Feb.) 2021]**

(a) Terminal B — H bonds have less *p*-character when compared to bridging bonds.

(b) The two B — H — B bonds are not of same length.

(c) All B — H — B angles are of 120°.

(d) Its fragment, BH_3 behaves as a Lewis base.

44. The secondary valency and the number of hydrogen bonded water molecule(s) in $CuSO_4 \cdot 5H_2O$, respectively, are : **[JEE Main (March) 2021]**

(a) 6 and 4 (b) 4 and 1 (c) 6 and 5 (d) 5 and 1

45. Given below are two statements : **[JEE Main (Feb.) 2021]**

Statement I : α and β forms of sulphur can change reversibly between themselves with slow heating or slow cooling.

Statement II : At room temperature the stable crystalline form of sulphur is monoclinic sulphur.

In the light of the above statements, choose the correct answer from the options given below.

(a) Statement I is false but Statement II is true.

(b) Both Statement I and Statement II are true.

(c) Statement I is true but Statement II is false.

(d) Both Statement I and Statement II are false.

46. Given below are two statements: one is labelled as Assertion A and the other is labelled as Reason R. **[JEE Main (Feb.) 2021]**

Assertion A : Dipole-dipole interactions are the only non-covalent interactions, resulting in hydrogen bond formation.

Reason R : Fluorine is the most electronegative element and hydrogen bonds in HF are symmetrical. In the light of the above statements, choose the most appropriate answer from the options given below.

(a) A is false but R is true

(b) Both A and R are true and R is the correct explanation of A

(c) A is true R is false

(d) Both A and R are true but R is not the correct explanation of A.

47. Given below are two statements : **[JEE Main (Feb.) 2021]**

Statement I : *o*-Nitrophenol is steam volatile due to intramolecular hydrogen bonding.

Statement II : *o*-Nitrophenol has high melting due to hydrogen bonding.

In the light of the above statements, choose the most appropriate answer from the options given below:

(a) Statement I is false but Statement II is true

(b) Both statement I and statement II are true

(c) Both statement I and statement II are false

(d) Statement I is true but statement II is false

48. The incorrect statement regarding the structure of C_{60} is : **[JEE Main (March) 2021]**
(a) The six-membered rings are fused to both six and five-membered rings.
(b) Each carbon atom forms three sigma bonds.
(c) The five-membered rings are fused only to six-membered rings.
(d) It contains 12 six-membered rings and 24 five-membered rings.

49. Arrange the following in the decreasing order of their covalent character :
[JEE Main (June) 2022 (I)]
(A) LiCl (B) NaCl (C) KCl (D) CsCl
Choose the most appropriate answer from the options given below :
(a) (A) > (C) > (B) > (D) (b) (B) > (A) > (C) > (D)
(c) (A) > (B) > (C) > (D) (d) (A) > (B) > (D) > (C)

50. Number of lone pair(s) of electrons on central atom and the shape of BrF_3 molecule respectively, are : **[JEE Main (June) 2022 (II)]**
(a) 0, triangular planar (b) 1, pyramidal
(c) 2, bent T-shape (d) 1, bent T-shape

51. Identify the incorrect statement for PCl_5 from the following : **[JEE Main (June) 2022 (II)]**
(a) In this molecule, orbitals of phosphorous are assumed to undergo sp^3d hybridization
(b) The geometry of PCl_5 is trigonal bipyramidal
(c) PCl_5 has two axial bonds stronger than three equatorial bonds
(d) The three equatorial bonds of PCl_5 lie in a plane

52. Consider the ions/molecule $O_2^+, O_2, O_2^-, O_2^{2-}$
For increasing bond order the correct option is : **[JEE Main (June) 2022 (I)]**
(a) $O_2^{2-} < O_2^- < O_2 < O_2^+$ (b) $O_2^- < O_2^{2-} < O_2 < O_2^+$
(c) $O_2 < O_2^{2-} < O_2^+ < O_2$ (d) $O_2^- < O_2^+ < O_2^{2-} < O_2$

53. Order of covalent bond ; **[JEE Main (Jan.) 2023 (I)]**
A. KF > KI ; LiF > KF
B. KF < KI ; LiF > KF
C. $SnCl_4 > SnCl_2$; CuCl > NaCl
D. LiF > KF ; CuCl < NaCl
E. KF < KI ; CuCl > NaCl
(a) C, E only (b) B, C only (c) B, C, E only (d) A, B only

54. For OF_2 molecule consider the following : **[JEE Main (Jan.) 2023 (I)]**
(A) Number of lone pairs on oxygen is 2.
(B) FOF angle is less than 104.5°.
(C) Oxidation state of O is –2.
(D) Molecule is bent 'V' shaped.
(E) Molecular geometry is linear.
Correct options are :
(a) C, D, E only (b) B, E, A only (c) A, C, D only (d) A, B, D only

55. In which of the following processes, the bond order increases and paramagnetic character changes to diamagnetic one? **[JEE Main (April) 2023 (I)]**
(a) $O_2 \longrightarrow O_2^-$ (b) $NO \longrightarrow NO^+$ (c) $N_2 \longrightarrow N_2^+$ (d) $O_2 \longrightarrow O_2^+$

56. Decreasing order of the hydrogen bonding in following forms of water is correctly represented by : **[JEE Main (Jan.) 2023 (I)]**

A. Liquid water B. Ice C. Impure water

(a) A = B > C (b) B > A > C (c) C > B > A (d) A > B > C

57. Match List-I with List-II : **[JEE Main (Jan.) 2024]**

List-I (Molecule)	**List-II (Shape)**
(A) BrF_5	(I) T-shape
(B) H_2O	(II) See-saw
(C) ClF_3	(III) Bent
(D) SF_4	(IV) Square pyramidal

(a) (A)–I, (B)–II, (C)–IV, (D)–III (b) (A)–II, (B)–I, (C)–III, (D)–IV

(c) (A)–III, (B)–IV, (C)–I, (D)–II (d) (A)–IV, (B)–III, (C)–I, (D)–II

58. Which of the following is least ionic? **[JEE Main (Jan.) 2024]**

(a) $BaCl_2$ (b) AgCl (c) KCl (d) $CoCl_2$

59. Given below are two statements : one is labelled as Assertion (A) and the other is labelled as Reason (R).

Assertion (A) : There is a considerable increase in covalent radius from N to P. However from As to Bi only a small increase in covalent radius is observed.

Reason (R) : Covalent and ionic radii in a particular oxidation state increases down the group.

In the light of the above statement, choose the most appropriate answer from the options given below : **[JEE Main (Jan.) 2024]**

(a) (A) is false but (R) is true

(b) Both (A) and (R) are true but (R) is not the correct explanation of (A)

(c) (A) is true but (R) is false

(d) Both (A) and (R) are true and (R) is the correct explanation of (A)

60. Arrange the following compounds in increasing order of their dipole moment :

HBr, H_2S, NF_3 and $CHCl_3$ **[JEE Main (Jan.) 2025]**

(a) $NF_3 < HBr < H_2S < CHCl_3$ (b) $HBr < H_2S < NF_3 < CHCl_3$

(c) $H_2S < HBr < NF_3 < CHCl_3$ (d) $CHCl_3 < NF_3 < HBr < H_2S$

61. Which of the following statement is true with respect to H_2O, NH_3 and CH_4?

A. The central atoms of all the molecules are sp^3 hybridized.

B. The H—O—H, H—N—H and H—C—H angles in the above molecules are 104.5°, 107.5° and 109.5° respectively.

C. The increasing order of dipole moment is $CH_4 < NH_3 < H_2O$.

D. Both H_2O and NH_3 are Lewis acids and CH_4 is a Lewis base.

E. A solution of NH_3 in H_2O is basic. In this solution NH_3 and H_2O act as Lowry-Bronsted acid and base respectively.

Choose the correct answer from the options given below : **[JEE Main (Jan.) 2025]**

(a) A, B and C only (b) C, D and E only

(c) A, D and E only (d) A, B, C and E only

62. Consider 'n' is the number of lone pair of electrons present in the equatorial position of the most stable structure of ClF_3. The ions from the following with 'n' number of unpaired electrons are :

A. V^{3+} B. Ti^{3+} C. Cu^{2+} D. Ni^{2+} E. Ti^{2+}

Choose the correct answer from the options given below : **[JEE Main (Jan.) 2025]**

(a) A and C only (b) A, D and E only (c) B and C only (d) B and D only

63. The molecules having square pyramidal geometry are : **[JEE Main (Jan.) 2025]**

(a) BrF_5 and $XeOF_4$ (b) SbF_5 and $XeOF_4$ (c) SbF_5 and PCl_5 (d) BrF_5 and PCl_5

64. The type of oxide formed by the element among Li, Na, Be, Mg, B and Al that has the least atomic radius is: **[JEE Main (Jan.) 2025]**

(a) A_2O_3 (b) AO_2 (c) AO (d) A_2O

B. Objective Questions

[One or more than one option(s) is/are correct]

1. According to molecular orbital theory, **[JEE (Advanced) 2016]**

(a) C_2^{2-} is expected to be diamagnetic

(b) O_2^{2+} is expected to have a longer bond length than O_2

(c) N_2^+ and N_2^- have the same bond order

(d) He_2^+ has the same energy as two isolated He atoms.

2. Each of the following options contains a set of four molecules. Identify the option(s) where all four molecules possesses permanent dipole moment at room temperature. **[JEE (Advanced) 2019]**

(a) NO_2, NH_3, $POCl_3$, CH_3Cl (b) BF_3, O_3, SF_6, XeF_6

(c) $BeCl_2$, CO_2, BCl_3, $CHCl_3$ (d) SO_2, C_6H_5Cl, H_2Se, BrF_5

3. In an experiment, m grams of a compound X (gas/liquid/solid) taken in a container is loaded in a balance as shown in (figure I) below. In the presence of a magnetic field, the pan with X is either deflected upwards (figure II), or deflected downwards (figure III), depending on the compound X. Identify the correct statement (s). **[JEE (Advanced) 2020]**

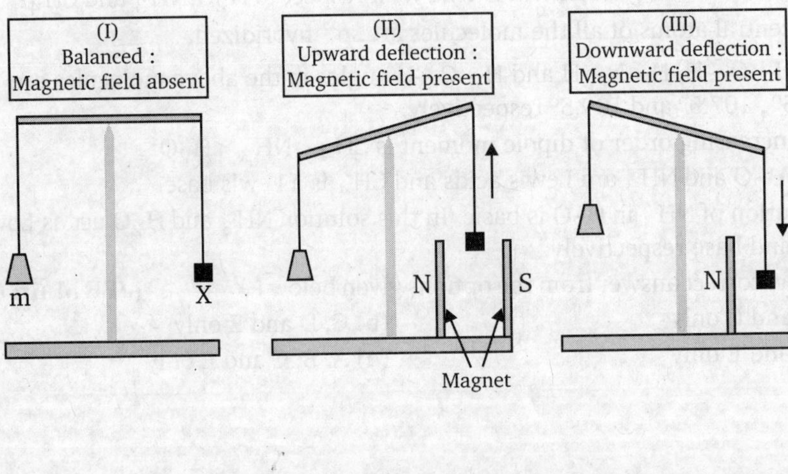

(a) If X is $H_2O(l)$, deflection of the pan is upwards.
(b) If X is $K_4[Fe(CN)_6](s)$, deflection of the pan is upwards.
(c) If X is $O_2(g)$, deflection of the pan is downwards.
(d) If X is $C_6H_6(l)$, deflection of the pan is downwards.

C. Matching Type Problems

1. Match of the diatomic molecules in **Column I** with its property/properties in **Column II**: **(IIT 2009)**

	Column I		Column II
(a)	B_2	(p)	Paramagnetic
(b)	N_2	(q)	Undergoes oxidation
(c)	O_2^-	(r)	Undergoes reduction
(d)	O_2	(s)	Bond order ≥ 2.
		(t)	Mixing of 's' and 'p' orbital

2. Match the orbital overlap figures shown in List-I with the description given in List-II and select the correct answer using the code given below the lists. **[JEE (Advanced) 2014]**

List-I		List-II
(A) 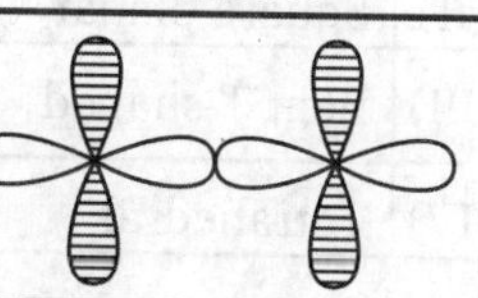		(p) p-$d\pi$ antibonding
(B) 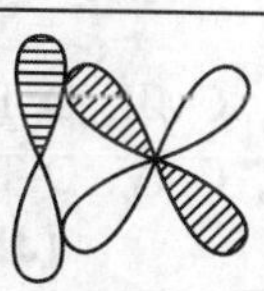		(q) d-$d\sigma$ bonding
(C) 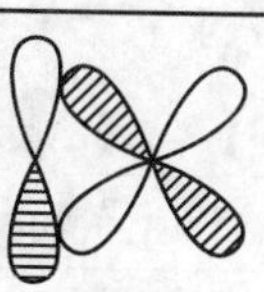		(r) p-$d\pi$ bonding
(D) 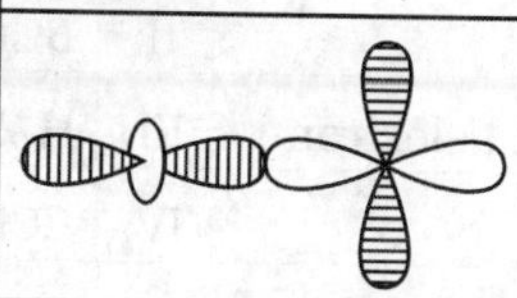		(s) d-$d\sigma$ antibonding

Code :

	A	B	C	D
(a)	q	p	r	s
(b)	s	r	p	q
(c)	q	r	p	s
(d)	s	p	r	q

3. Match List-I with List-II : **[JEE Main (April) 2023 (I)]**

	List-I (Species)		List-II (Geometry/Shape)
(A)	H_3O^+	(I)	Tetrahedral
(B)	Acetylide anion	(II)	Linear
(C)	NH_4^+	(III)	Pyramidal
(D)	ClO_2^-	(IV)	Bent

Choose the correct answer from the options given below :

(a) A–III, B–II, C–I, D–IV
(b) A–III, B–I, C–II, D–IV
(c) A–III, B–IV, C–I, D–II
(d) A–III, B–IV, C–II, D–I

4. Match List-I with List-II : **[JEE Main (Jan.) 2023 (I)]**

	List-I		List-II
(A)	XeF_4	(I)	See-saw
(B)	SF_4	(II)	Square planar
(C)	NH_4^+	(III)	Bent T-shaped
(D)	BrF_3	(IV)	Tetrahedral

Choose the correct answer from the options given below :

(a) A–IV, B–III, C–II, D–I
(b) A–II, B–I, C–III, D–IV
(c) A–IV, B–I, C–II, D–III
(d) A–II, B–I, C–IV, D–III

5. Match the **LIST-I** with **LIST-II.** **[JEE Main (Jan.) 2025]**

	LIST-I (Classification of molecules based on octet rule)		LIST-II (Example)
A.	Molecules obeying octet rule	I.	NO, NO_2
B.	Molecules with incomplete octet	II.	$BCl_3, AlCl_3$
C.	Molecules with incomplete octet with odd electron	III.	H_2SO_4, PCl_5
D.	Molecules with expanded octet	IV.	CCl_4, CO_2

Choose the correct answer from the options given below:

(a) A-IV, B-II, C-I, D-III
(b) A-III, B-II, C-I, D-IV
(c) A-IV, B-I, C-III, D-II
(d) A-II, B-IV, C-III, D-I

6. Match **List-I** with **List-II.** **[JEE Main (Jan.) 2025]**

List-I (Complex)		List-II (Hybridisation of central metal ion)	
(A)	$[CoF_6]^{3-}$	(I)	d^2sp^3
(B)	$[NiCl_4]^{2-}$	(II)	sp^3
(C)	$[Co(NH_3)_6]^{3+}$	(III)	sp^3d^2
(D)	$[Ni(CN)_4]^{2-}$	(IV)	dsp^2

Choose the correct answer from the options given below :

(a) (A)-(I), B-(IV), (C)-(III), (D)-(II)
(b) (A)-(III), B-(II), (C)-(I), (D)-(IV)
(c) (A)-(I), B-(II), (C)-(III), (D)-(IV)
(d) (A)-(III), B-(IV), (C)-(I), (D)-(II)

D. Statement and Explanation Type Problems

Read the following question and answer as per the direction given below :

(a) Statement 1 is true; Statement 2 is true; Statement 2 is the correct explanation of Statement 1.
(b) Statement 1 is true; Statement 2 is true; Statement 2 is not the correct explanation of Statement 1.
(c) Statement 1 is true; Statement 2 is false.
(d) Statement 1 is false; Statement 2 is true

1. Statement 1 : Band gap in germanium is small.
Statement 2: The energy spread of each germanium atomic energy level is infinitesimally small. **(IIT 2007)**

E. Integer Type Problems

1. Based on VSEPR theory, the number of 90° F — Br — F angles in BrF_5 is : **(IIT 2010)**

2. A list of species having the formula XZ_4 is given below:
$XeF_4, SF_4, SiF_4, BF_4^-, BrF_4^-, [Cu(NH_3)_4]^{2+}, [FeCl_4]^{2-}, [CoCl_4]^{2-}$ and $[PtCl_4]^{2-}$.
Defining shape on the basis of the location of X and Z atoms, the total number of species having a square planar shape is .... **[JEE (Advanced) 2014]**

3. The total number of lone pairs of electrons in N_2O_3 is : **[JEE (Advanced) 2015]**

4. Among the triatomic molecules/ions, $BeCl_2, N_3^-, N_2O, NO_2^+, O_3, SCl_2, ICl_2^-, I_3^-$ and XeF_2, the total number of linear molecule(s)/ion(s) where the hybridization of the central atom does not have contribution from the d-orbital(s) is :
[Atomic number : S = 16, Cl = 17, I = 53 and Xe = 54] **[JEE (Advanced) 2015]**

5. The sum of the number of lone pairs of electrons on each central atom in the following species is: $[TeBr_6]^{2-}$, $[BrF_2]^+$, SNF_3 and $[XeF_3]^-$

(Atomic numbers: N = 7, F = 9, S = 16, Br = 35, Te = 52, Xe = 54) **[JEE (Advanced) 2017]**

6. Among $H_2, He_2^+, Li_2, Be_2, B_2, C_2, N_2, O_2^-$ and F_2 the number of diamagnetic species is : (Atomic numbers: H=1, He=2, Li=3, Be=4, B=5, C=6, N=7, O=8, F=9)

[JEE (Advanced) 2017]

7. Among B_2H_6, $B_3N_3H_6$, N_2O, N_2O_4, $H_2S_2O_3$ and $H_2S_2O_8$, the total number of molecules containing covalent bond between two atoms of the same kind is ___.

[JEE (Advanced) 2019]

8. Consider the kinetic data given in the following table for the reaction, $A + B + C \rightarrow$ product.

Experiment No.	[A] (mol dm^{-3})	[B] (mol dm^{-3})	[C] (mol dm^{-3})	Rate of reactions (mol dm^{-3}s^{-1})
1	0.2	0.1	0.1	6.0×10^{-5}
2	0.2	0.2	0.1	6.0×10^{-5}
3	0.2	0.1	0.2	1.2×10^{-4}
4	0.3	0.1	0.1	9.0×10^{-5}

The rate of the reaction for $[A] = 0.15$ mol dm^{-3}, $[B] = 0.25$ mol dm^{-3} and $[C] = 0.15$ mol dm^{-3} is found to be $Y \times 10^{-5}$ mol dm^{-3}s^{-1}. The value of Y is ___. **[JEE (Advanced) 2019]**

9. Consider the following compounds in the liquid form :

$$O_2, HF, H_2O, NH_3, H_2O_2, CCl_4, CHCl_3, C_6H_6, C_6H_5Cl.$$

When a charged comb is brought near their flowing stream, how many of them show deflection as per the following figure? **[JEE (Advanced) 2020]**

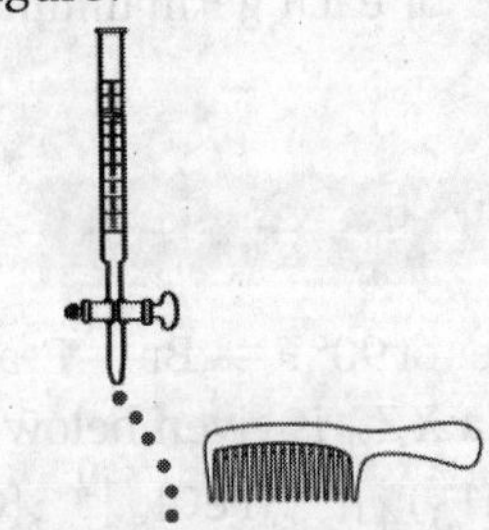

10. The figure below is the plot of potential energy versus internuclear distance (d) of H_2 molecule in the electronic ground state. What is the value of the net potential energy E_0 (as indicated in the figure) in kJ mol^{-1}, for $d = d_0$ at which the electron-electron repulsion and the nucleus-nucleus repulsion energies are absent? As reference, the potential energy of H atom is taken as zero when its electron and the nucleus are infinitely far apart.

Use Avogadro constant as 6.023×10^{23} mol^{-1}. **[JEE (Advanced) 2020]**

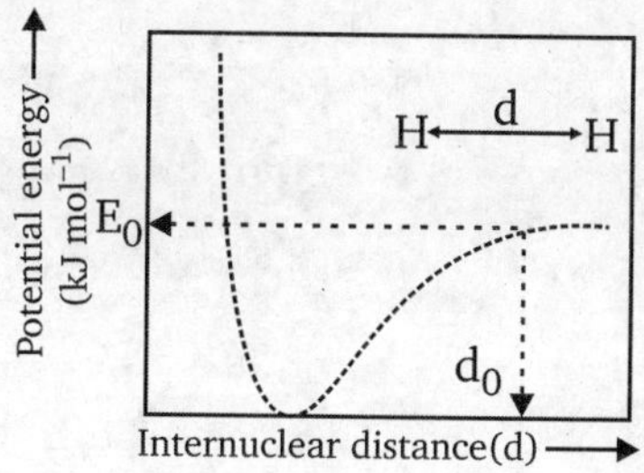

11. The sum of the total number of bonds between chromium and oxygen atoms in chromate and dichromate ions is: **[JEE Main (Jan.) 2020]**

12. The number of species below that have two lone pairs of electrons in their central atom is ____ (Round off to the nearest integer)
$SF_4, BF_4^-, ClF_3, AsF_3, PCl_5, BrF_5, XeF_4, SF_6$ **[JEE Main (March) 2021]**

13. A xenon compound '*A*' upon partial hydrolysis gives XeO_2F_2. The number of lone pair of electrons present in compound *A* is ______ (Round off to the nearest integer) **[JEE Main (March) 2021]**

14. *AX* is a covalent diatomic molecule where *A* and *X* are second row elements of periodic table. Based on molecular orbital theory, the bond order of *AX* is 2.5. The total number of electrons in *AX* is.......(Round off to the nearest integer). **[JEE Main (March) 2021]**

15. The hybridization of P exhibited in PF_5 is sp^xd^y. The value of *y* is ______. **[JEE Main (June) 2022 (I)]**

16. The number of molecules from the following which contain only two lone pair of electrons is ______.
$H_2O, N_2, CO, XeF_4, NH_3, NO, CO_2, F_2$ **[JEE Main (April) 2023 (II)]**

17. The maximum number of lone pairs of electrons on the central atom from the following species is ______.
ClO_3^-, XeF_4, SF_4 and I_3^- **[JEE Main (April) 2023 (II)]**

18. The total number of anti bonding molecular orbitals, formed from 2*s* and 2*p* atomic orbitals in a diatomic molecule is ______. **[JEE Main (Jan.) 2024]**

19. The number of molecules/ion/s having trigonal bipyramidal shape is ....
$PF_5, BrF_5, PCl_5, [PtCl_4]^{2-}, BF_3, Fe(CO)_5$ **[JEE Main (Jan.) 2024]**

20. Sum of bond order of CO and NO^+ is ______. **[JEE Main (Jan.) 2024]**

21. The number of molecules/ions that show linear geometry among the following is ....
SO_2, $BeCl_2, CO_2, N_3^-$, $NO_2, F_2O, XeF_2, NO_2^+$, I_3^-, O_3 **[JEE Main (Jan.) 2025]**

22. Total number of non-bonded electrons present in NO_2^- ion based on Lewis theory is .... **[JEE Main (Jan.) 2025]**

Answers

[A] Objective Questions (Only one option is correct)

1. (c) **2.** (d) **3.** (c) **4.** (c) **5.** (b) **6.** (a) **7.** (d) **8.** (c) **9.** (b) **10.** (d)
11. (a) **12.** (a) **13.** (b) **14.** (b) **15.** (a) **16.** (d) **17.** (d) **18.** (a) **19.** (c) **20.** (d)
21. (b) **22.** (d) **23.** (b) **24.** (b) **25.** (a) **26.** (c) **27.** (c) **28.** (c) **29.** (b) **30.** (c)
31. (c) **32.** (d) **33.** (c) **34.** (c) **35.** (d) **36.** (c) **37.** (b) **38.** (d) **39.** (c) **40.** (d)
41. (d) **42.** (c) **43.** (d) **44.** (b) **45.** (c) **46.** (a) **47.** (d) **48.** (d) **49.** (c) **50.** (c)
51. (c) **52.** (a) **53.** (c) **54.** (d) **55.** (b) **56.** (b) **57.** (d) **58.** (b) **59.** (b) **60.** (a)
61. (a) **62.** (b) **63.** (a) **64.** (a)

[B] Objective Questions (One or more than one option(s) is/are correct)

1. (a, c) **2.** (a, d) **3.** (a, b, c)

[C] Matching Type Problems

1. (a) – p, r, t; (b) – s, t; (c) – p, q; (d) – p, q, s
2. (c) (A) – q; (B) – r; (C) – p; (D) – s
3. **(a)** A–III, B–II, C–I, D–IV
4. **(d)** A–II, B–I, C–IV, D–III
5. **(a)** A-IV, B-II, C-I, D-III
6. **(b)** (A)-(III), B-(II), (C)-(I), (D)-(IV)

[D] Statement and Explanation Type Problems

1. (b)

[E] Integer Type Problems

1. (0) **2.** (4) **3.** (8) **4.** (4) **5.** (6)
6. (6) **7.** (4) **8.** (6.75) **9.** (6) **10.** (5246.49)
11. (12) **12.** (2) **13.** (19) **14.** (15) **15.** (1)
16. (4) **17.** (3) **18.** (4) **19.** (3) **20.** (6)
21. (6) **22.** (12)

Hints and Solutions

A. Objective Questions

[Only one option is correct]

1. (c) C — C bond is of considerable strength so (a) and (b) can not be the right choice. (d) is not correct as this value is quite large.

2. (d) 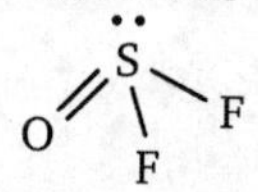

SO_3 is trigonal planar, BrF_3 is bent T-shaped, SiO_3^{2-} is tetrahedral.

3. (c) O_2 is expected to diamagnetic in nature but actually it is paramagnetic.

4. (c) H_2^{2+} and H_2 does not exist.

5. (b) B.O. of $Li_2^+ = 0.5$;   B.O. of $Li_2^- = 0.5$

Hence stability order $= Li_2^- < Li_2^+ < Li_2$

6. (a) Nitric oxide is paramagnetic in the gaseous state as it has one unpaired electron in its outermost shell. The electronic configuration of NO is

$$\sigma_{1s}^2\sigma^{*2}_{1s}\,\sigma^2_{2s}\,\sigma^{*2}_{2s}\,\sigma^2_{2p_z}\,\pi^2_{2p_x} = \pi^2_{2p_y}\,\pi^{*1}_{2p_x}$$

However, it dimerises at low temperature to become diamagnetic.

$$2NO \rightleftharpoons N_2O_2$$

Its bond order is 2.5 and it combines with O_2 to give nitrogen dioxide.

7. (d) 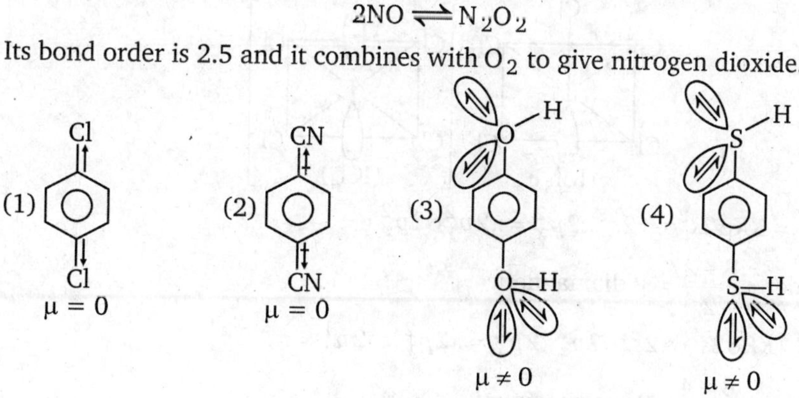

8. (c) $\sigma 1s^2, \sigma^* 1s^2, \sigma 2s^2, \sigma^* 2s^2, \sigma 2p_z^2, \begin{matrix}\pi 2p_x^1 \\ \pi 2p_y^1\end{matrix}, \begin{matrix}\pi^* 2p_x^0 \\ \pi^* 2p_y^0\end{matrix}, \sigma^* 2p_z^0$

(paramagnetic)

Only C_2 is paramagnetic.

9. (b) Dipole-dipole interaction (hydrogen bonding) is proportional to $1/r^3$, where r is the distance between the polar molecules.

10. (d) (a) $O_2(16): \sigma 1s^2 \sigma * 1s^2 \sigma 2s^2 \sigma * 2s^2 \sigma 2p_z^2\ \pi 2p_x^2 = \pi 2p_y^2\ \pi^* 2p_x^1 = \pi^* 2p_y^1$

Number of unpaired electrons $=1$ (paramagnetic)

(b) $B_2(10): \sigma 1s^2 \sigma^* 1s^2 \sigma 2s^2 \sigma^* 2s^2\ \pi 2p_x^1 = \pi 2p_y^1$

Number of unpaired electrons $=2$ (paramagnetic)

(c) $NO(15): \sigma 1s^2 \sigma^* 1s^2 \sigma 2s^2 \sigma^* 2s^2 \sigma 2p_z^2\ \pi 2p_x^2 = \pi 2p_y^2\ ^* 2p_x^1$

Number of unpaired electrons $=1$ (paramagnetic)

(d) $CO(14): \sigma 1s^2 \sigma^* 1s^2 \sigma 2s^2 \sigma^* 2s^2\ \pi 2p_x^2 = \pi 2p_y^2\ \sigma 2p_z^2$

Number of unpaired electron $=0$ (diamagnetic)

11. (a) I_3^- is - sp^3d hybridised

- linear shape

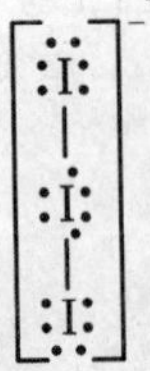

Total number of lone pair of electron=9

12. (a) KCl exist as K^+ and Cl^-

13. (b) H_2^{2-} does not exist as Bond order is zero.

Electronic configuration of $H_2^{2-}: \sigma_{1s}^2 \sigma_{1s}^{*2}$

$$B.O. = \frac{2-2}{2} = 0$$

14. (b) Shape of ICl_5 is square pyramidal and ICl_4^- is square planar.

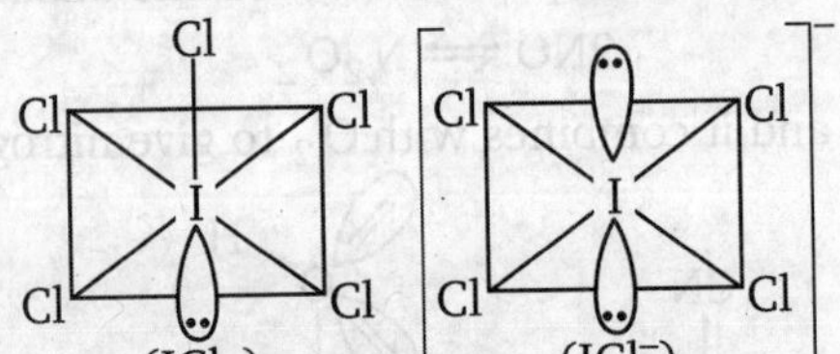

(ICl_5) (ICl_4^-)

15. (a) $C_2^{2-}(14) := KK\sigma 2s^2 \sigma * 2s^2 \pi 2p_x^2 = \pi 2p_y^2 \sigma 2p_z^2$

$$\text{Bond order} = \frac{8-2}{2} = 3;\ \text{diamagnetic}$$

$O_2(16) := KK\sigma 2s^2 \sigma * 2s^2 \sigma 2p_z^2 \pi 2p_x^2 = \pi 2p_y^2\ \pi * 2p_x^1 = \pi * 2p_y^1$

$$\text{Bond order} = \frac{8-4}{2} = 2;\ \text{paramagnetic}$$

$N_2^{2-}(16) = KK\sigma 2s^2 \sigma * 2s^2 \pi 2p_x^2 = \pi 2p_y^2 \sigma 2p_z^2 \pi * 2p_x^1 = \pi * 2p_y^1$

Bond order $= \dfrac{8-4}{2} = 2$; paramagnetic

$O_2^{2-}(18) := KK\sigma * 2s^2\sigma * 2s^2\sigma 2p_z^2\pi 2p_x^2 = \pi 2p_y^2\pi * 2p_x^2 = \pi * 2p_y^2$

Bond order $= \dfrac{8-6}{2} = 1.0$; diamagnetic

As bond order $\propto \dfrac{1}{\text{Bond length}}$, therefore, C_2^{2-} has shortest bond length among the given molecules.

16. (d) Since, during formation of *A*—*B* bond maximum energy is released. So *A*—*B* bond will be strongest in all given bonds.

❖ Bond strength $\propto \dfrac{1}{\text{Bond length}}$

Also bond length of *A*—*B* bond will be minimum.

❖ Electronegativity of *B* atom will be maximum since it forms strongest bond.

17. (d) As atoms come closer potential energy decreases due to attractive forces. It is minimum when a bond is formed after that potential energy starts rising due to repulsive interaction of electronic change.

18. (a) Bond energy $\propto \dfrac{1}{\text{Bond length}}$

Order of bond length : C — I > C — Br > C — Cl > C — F

(Since down the group size of atom increases)

Order of bond energy : C — F > C — Cl > C — Br > C — I

19. (c) XeF_6 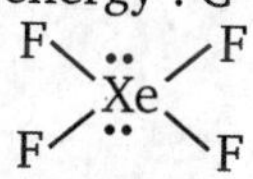 sp^3d^2

$[CrF_6]^{3-}$  d^2sp^3

$[Ni(CN)_4]^{2-}$ 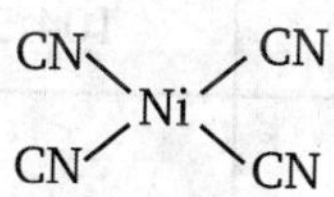 dsp^2

BrF_5 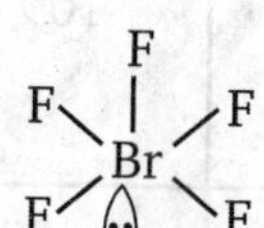 sp^3d^2

20. (d) N_2 is used as inert diluent for chemicals. Due to triple bond, its bond energy is high, so it is chemically inert.

21. (b) AB_4 molecule is a polar molecule.

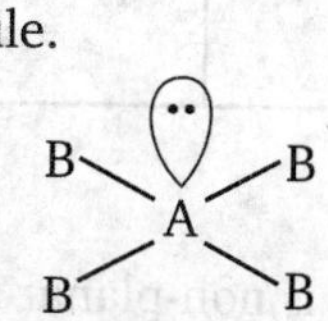

Square pyramidal will be polar because of Non-symmetrical/Non-equivalent structure.

22. (d) [XeF_5^-]

Hybridisation = sp^3d^3

Bond pairs = 5 $\Rightarrow$ Lone pairs = 2

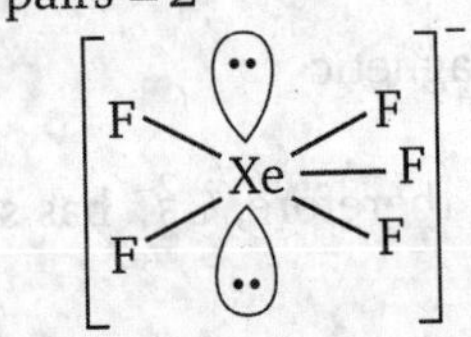

Shape → pentagonal planar

[XeO_3F_2]

Hybridisation = sp^3d

Bond pair = 5 $\Rightarrow$ Lone pair = 0

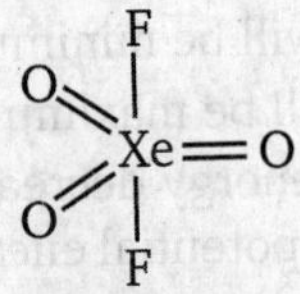

Shape → trigonal bipyramidal.

23. (b) In SF_4 → hybridisation = sp^3d

Bond pair = 4; Lone pair = 1

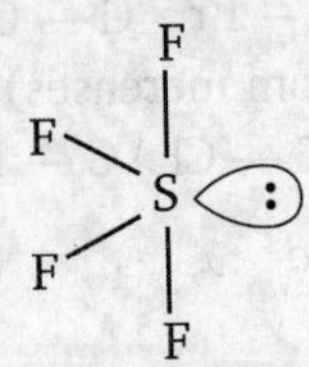

Shape: trigonal bipyramidal. (Including lone pairs)

24. (b)

Compound	Hybridisation	Structure	Bond angle
H_2O	sp^3	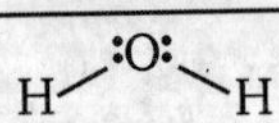	104°5'
CH_4	sp^3	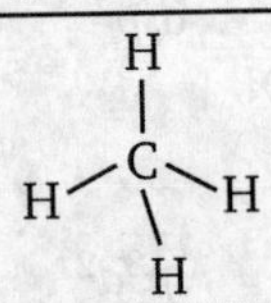	109°28'
H_2S	sp^3	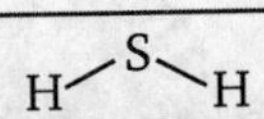	92°
NH_3	sp^3	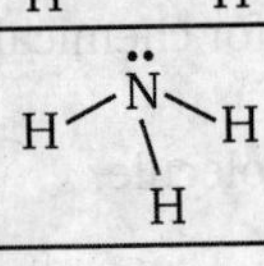	107°

25. (a) 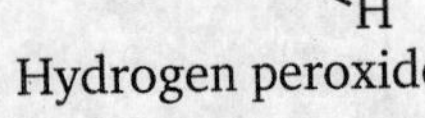

Hydrogen peroxide in pure state is non-planar and almost colourless liquid.

26. (c) CCl_4 and CH_4 are symmetrical molecules have zero dipole moments ($\mu = 0$). While $CHCl_3$ is non-symmetrical have non-zero dipole moment ($\mu \neq 0$).

27. (c) Order of strength is ion-ion > ion-dipole > dipole-dipole.

28. (c) Boiling point order : $Z > Y > X$

Higher is the b.p of liquid, more will be intermolecular interaction between molecules. So, order of intermolecular interaction is: $Z > Y > X$.

29. (b) H_2O has higher boiling point than H_2S due to strong intermolecular H-bonding in H_2O so, b.p. of H_2O is 373 K and b.p of H_2S is 213 K.

30. (c) PCl_5 exist as $[PCl_4]^+ [PCl_6]^-$ in solid state.

Structure of PCl_4^+ is:

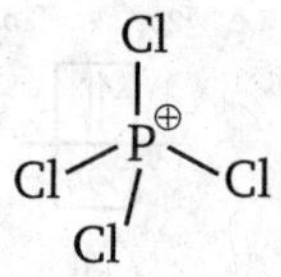

sp^3-hybridisation
(Tetrahedral)

Structure of PCl_6^- is:

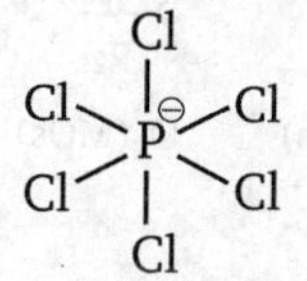

sp^3d^2-hybridisation
(Octahedral)

31. (c) $BeSO_4$ and $MgSO_4$ in water

For 2nd group sulphate down the group solubilities will decrease.

$$BeSO_4 > MgSO_4$$

Down the group radius of cation increases so decrease in hydration energy.
Decrease in hydration energy will lead to decrease in solubility.

Although both are highly soluble in water.

32. (d)

Species	**Bond order**
NO	2.5
NO^+	3
NO^{2+}	2.5
NO^-	2

Since, NO^- have minimum bond order, it will have minimum bond strength (bond energy).

33. (c) Alkali metal atoms always exist in +1 oxidation state in its compound.

34. (c) According to molecular orbital theory, configuration of CN^- is

$$\sigma_{1s}^2 \sigma_{1s}^{*2} \sigma_{2s}^2 \sigma_{2s}^{*2} \pi_{2p_x}^2 = \pi_{2p_y}^2, \sigma_{2p_x}^2$$

$$\text{Bond order} = \frac{1}{2}(N_b - N_a) = \frac{1}{2}(10 - 4) = 3$$

35. (d) $\mu = \sqrt{n(n+2)}$ B.M.

$\mu = 1.73$ B.M. $n = 1$

MOT diagram of O_2

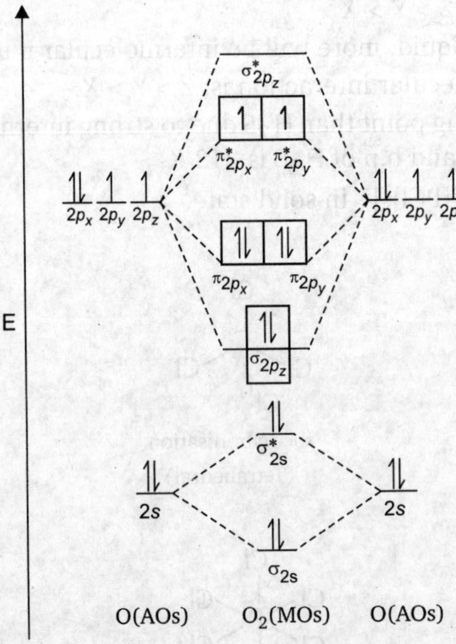

$n = 2$ for O_2

- O_2^- and O_2^+ both will have one unpaired electron
- $n = 1$ So $\mu = \sqrt{3} = 1.73$ B.M.

36. (c) Bond dissociation energy of halogen is $Cl_2 > Br_2 > F_2 > I_2$.
Bond dissociation enthalpy decreases down the group, fluorine (F_2) is exception due to high inter-electronic repulsion of lone pair.

37. (b) Isostructural species have same structures.

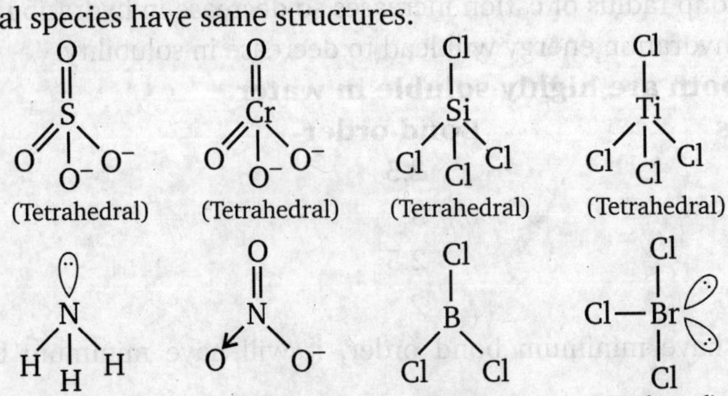

(Tetrahedral) (Tetrahedral) (Tetrahedral) (Tetrahedral)

(Trigonal pyramidal) (Trigonal planar) (Trigonal planar) (T-Shaped)

38. (d) Structure of I_3^-

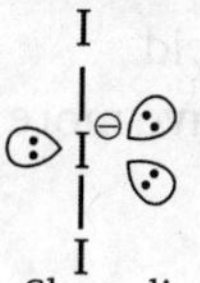

(sp^3d-hybridisation)

Shape-linear

Bond angle-180°.

39. (c)

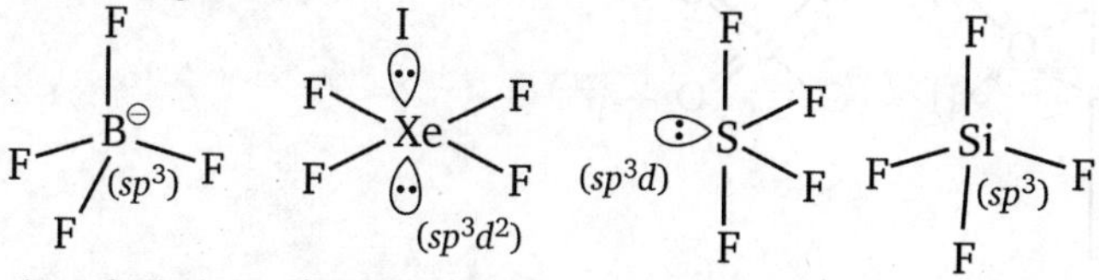

(Tetrahedral) (Square planar) (See-saw) (Tetrahedral)

❖ In SF_4 axial and equatorial bond length are different.

40. (d) Isomorphism: Two or more different substance having similar crystal structure.

❖ Isomorphous substance generally have similar chemical formulas.

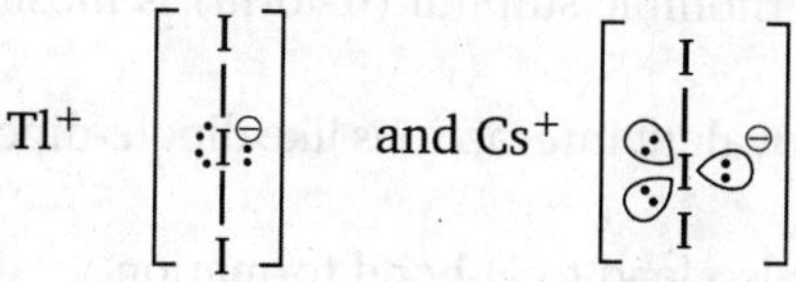

sp^3d sp^3d

Same crystal structure.

$Tl = 1s^2 2s^2 2p^6 3s^2 3p^6 4s^2 3d^{10} 4p^6 5s^2 4d^{10} 5p^6 6s^2 4f^{14} 5d^{10} 6p^1$

Both statements are correct but R is not correct explanation.

41. (d)

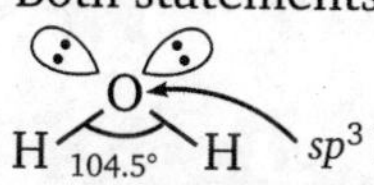

sp^3 hybridisation

Bond angle for tetrahedral structure should be 109.5° but as lone pairs around central oxygen repel more than bond pair, so bond angle is reduced by 5°.

42. (c)

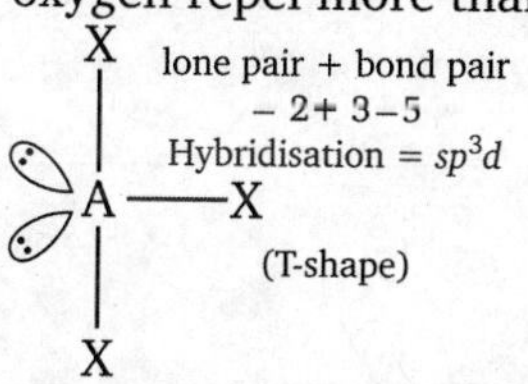

lone pair + bond pair

= 2+ 3–5

Hybridisation = sp^3d

(T-shape)

Molecule will be T-shaped.

43. (d)

Bridge B—H bond

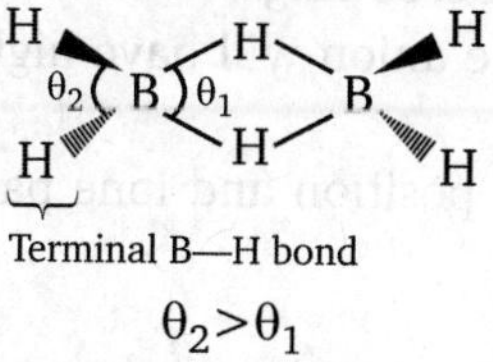

Terminal B—H bond

$\theta_2 > \theta_1$

❖ Both B — H — B bond are of same length.

❖ B — H — B bond angle are not 120°.

BH_3 have an empty *p*-orbital so it act as Lewis acid.

❖ Since, bridge B — H bond lengthen so it will have more *p*-character and terminal B — H bond will have less *p*-character.

44. (b) 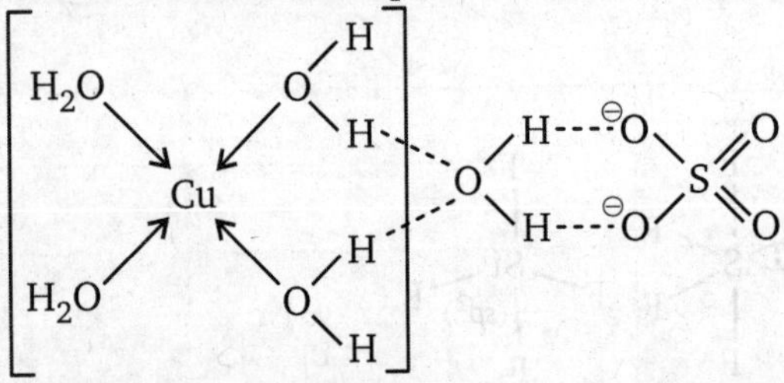

4 secondary valency

1 water molecule is hydrogen bonded.

45. (c) $S_{\text{Rhombic}}\ (\alpha\text{-form}) \underset{\text{cooling}}{\overset{\text{Heating}}{\rightleftharpoons}} S_{\text{Monoclinic}}\ (\beta\text{-form})$

At room temperature rhombic sulphur (α-form) is most stable form, so statement (II) is incorrect.

46. (a) It is correct that non-covalent interactions like dipole-dipole interaction can lead to H-bond formation.

❖ But ion dipole can also lead to H-bond formation.

e.g., $K^+F^{\ominus}$ - - - H — F strongest H-bond

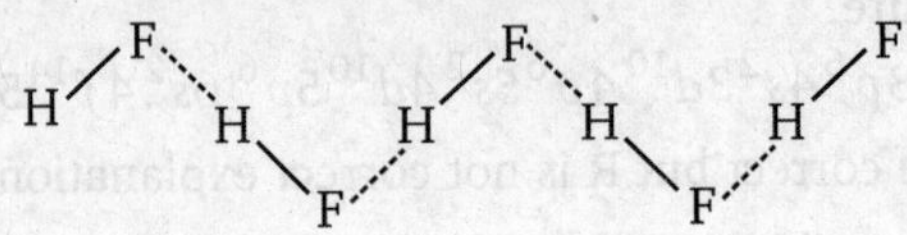

Symmetrical H-bonds are present.

Assertion is incorrect.

Reason is correct.

47. (d) 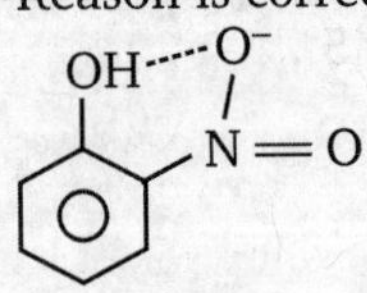

o-nitrophenol

is steam volatile due to intramolecular H-bonding.

❖ But it will have low melting point due to less packing efficiency.

Statement (I) is correct statement (II) is false.

48. (d) In C_{60} there are 12 five membered rings and 20 six membered rings.

49. (c) According to Fayan's rule, small size cation and large size anion will have high covalent character. So the order is : LiCl > NaCl > KCl > CsCl.

50. (c) As Fluorine is more electronegative it occupies the axial position and lone pairs occupy equatorial position.

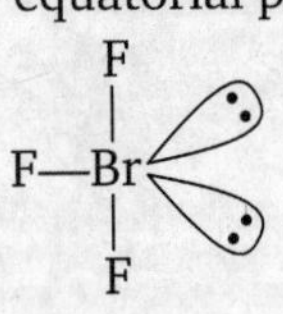

Steric no. = 5 (sp^3d), lone pair = 2

Bent T-shape.

51. (c) The sp^3d hybridization in PCl_5 is a combination of $[sp^2 + pd]$ mixing.
sp^2-hybrid orbitals form equitorial bonds and pd hybrid orbitals forms axial bonds. Hence, axial bonds are longer and weaker than equatorial bonds.

52. (a)

Ion/molecule	Number of e^- in BMO	Number of e^- in ABMO	Bond order
O_2^+	10	5	2.5
O_2	10	6	2
O_2^-	10	7	1.5
O_2^{2-}	10	8	1

Bond order : $O_2^{2-} < O_2^- < O_2 < O_2^+$

53. (c) According to Fajan's rule, covalent character is higher for large size anion and small size cation.

A. KF > KI — incorrect, LiF > KF – correct
B. KF < KI — correct, LiF > KF – correct
C. $SnCl_4 > SnCl_2$ — correct, CuCl > NaCl – correct
D. LiF > KF — correct, CuCl < NaCl – correct
E. KF < KI — correct, CuCl > NaCl – correct

54. (d)

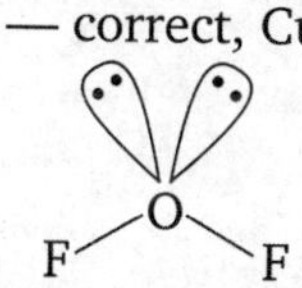

Molecule is bent 'v' shaped and oxidation state of oxygen is (+2).

55. (b) NO is paramagnetic with BO = 2.5, NO^+ is diamagnetic with BO = 3.

56. (b) Ice > Liquid water > Impure water
Due to impurity extent of H-bonding decreases.

57. (d)

BrF_5

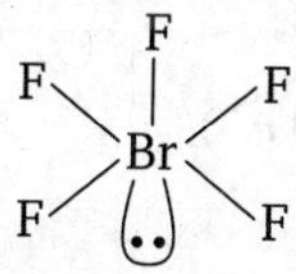

Square pyramidal

H_2O

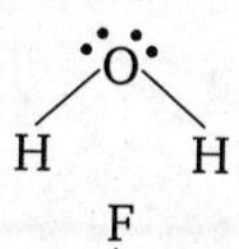

Bent

ClF_3

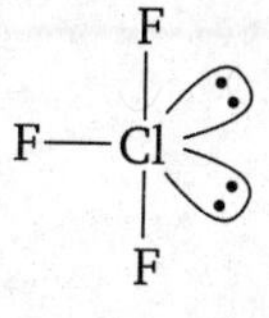

T-shape

SF_4 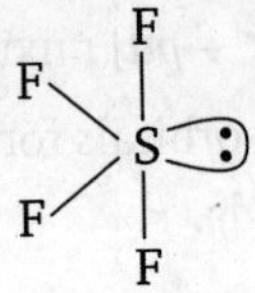 See-saw

58. (b) $AgCl < CoCl_2 < BaCl_2 < KCl$ (ionic character)

Reason : Ag^+ has pseudo inert gas configuration.

59. (b) According to NCERT,

Statement-I : Factual data,

Statement-II is true.

But correct explanation is presence of completely filled *d* and *f*-orbital of heavier members.

60. (a) Increasing order of dipole moment

$$\underset{\mu = 0.24\text{ D}}{NF_3} < \underset{0.79\text{ D}}{HBr} < \underset{0.95\text{ D}}{H_2S} < \underset{1.04\text{ D}}{CHCl_3}$$

It is NCERT data based.

61. (a) 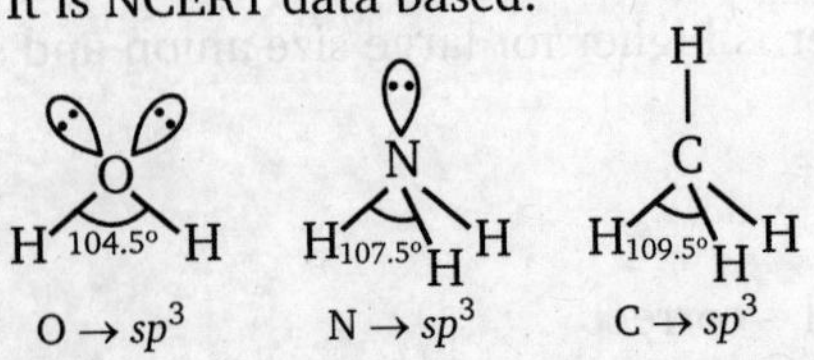

Dipole moment $H_2O > NH_3 > CH_4$

H_2O and NH_3 are Lewis Bases.

NH_3 act as Lowry-Bronsted base

Hence, A, B and C are correct.

62. (b) ClF_3

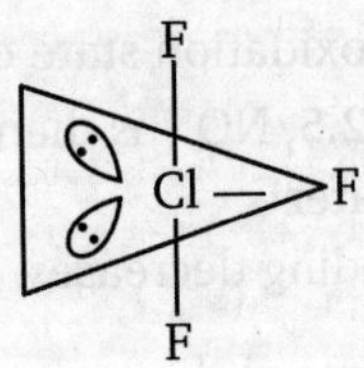

$n = 2$ (No. of lone pair present in equatorial plane)

	(Unpaired e^-)
(A) V^{3+}: $[Ar]3d^2$	2
(B) Ti^{3+}: $[Ar]3d^1$	1
(C) Cu^{2+}: $[Ar]3d^9$	1
(D) Ni^{2+}: $[Ar]3d^8$	2
(E) Ti^{2+}: $[Ar]3d^2$	2

63. (a)

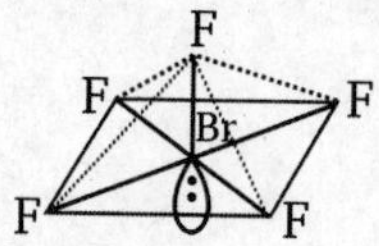

Square pyramidal

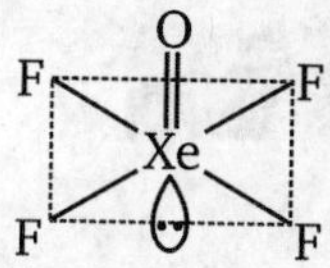

Square pyramidal

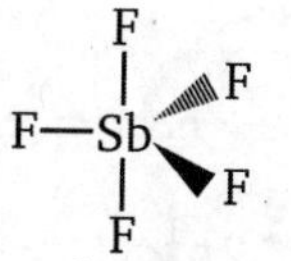

Trigonal Bipyramidal

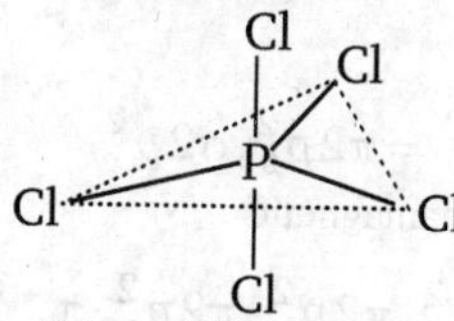

Trigonal Bipyramidal

BrF_5 : Square pyramidal
$XeOF_4$: Square pyramidal
SbF_5: Trigonal bipyramidal
PCl_5: Trigonal bipyramidal

B. Objective Questions

[One or more than one option(s) is/are correct]

1. (a, c)

(a) is true because all electrons are paired.
(b) is false. Bond length of O_2^{2+} < bond length of O_2.
(c) is true because both have same bond order (2.5).
(d) is false.

2. (a, d)

Symmetrical molecules like BF_3, SF_6, $BeCl_2$, CO_2 has zero dipole moment even though they contain polar bonds. Unsymmetrical molecules like $POCl_3$ (tetrahedral), CH_3Cl (tetrahedral), $CHCl_3$ (tetrahedral), XeF_6 (distorted octahedral), NH_3 (pyramidal), SO_2 (bent), H_2Se (V-shaped), NO_2 (V-shaped), O_3 (V-shaped), BrF_5 (square pyramidal) always have net value of dipole moment.

3. (a, b, c)

Paramagnetic compound (X) is attracted towards magnetic field and the pan is deflected downwards.
While the diamagnetic compound (X) is repelled by magnetic field and pan is deflected upward.

(a) $X \Rightarrow H_2O \rightarrow$ Diamagnetic (correct)
(b) $X \Rightarrow K_4[Fe(CN)_6](s) \rightarrow$ Diamagnetic (correct)
Here Fe^{2+} + Strong field ligand$\rightarrow 3d^6 \Rightarrow [t_{2g}^6, e_g^0]$
(c) $X \Rightarrow O_2 \rightarrow$ Paramagnetic (correct)
Here $O_2(g)$ is paramagnetic due to two-unpaired electrons present in π^* (antibonding orbitals).
(d) $X \Rightarrow C_6H_6(l) \rightarrow$ Diamagnetic (in correct)
It is due to presence of 0 unpaired electrons.

C. Matching Type Problems

1. (a) – p,r,t; (b) – s,t; (c) – p,q; (d) – p,q,s

(a) B_2: $\sigma 1s^2, \sigma * 1s^2, \sigma 2s^2, \sigma * 2s^2, [\pi 2p_y^1 = \pi 2p_z^1]$

Bond order = 1 Paramagnetic

(b) N_2: $\sigma 1s^2, \sigma * 1s^2, \sigma 2s^2, \sigma * 2s^2, \pi 2p_y^2 = \pi 2p_z^2, \sigma 2p_x^2$

Bond order = 3 Diamagnetic

(c) O_2^-: $\sigma 1s^2, \sigma * 1s^2, \sigma 2s^2, \sigma * 2s^2, \sigma 2p_x^2, \pi 2p_y^2, \pi 2p_z^2, \pi * 2p_y^2, \pi * 2p_z^1$

Bond order = 3/2 Paramagnetic

(d) O_2:

$\sigma 1s^2, \sigma * 1s^2, \sigma 2s^2, \sigma * 2s^2, \sigma 2p_x^2, \pi 2p_y^2, \pi 2p_z^2, \pi * 2p_y^1, \pi * 2p_z^1$

Bond order = 2 Paramagnetic

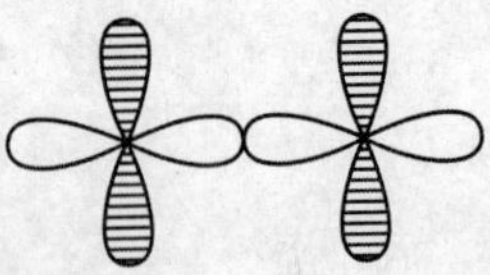

2. **(A) q :** ⟶ Both the *d*-orbitals show axial overlapping in same phase.
So it is *d*-*d* σ bonding.

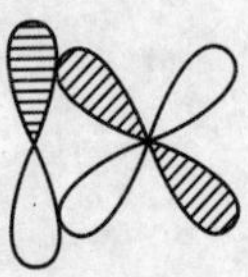

(B) r: ⟶ Both *p* and *d*-orbitals show lateral overlapping in same phase, so it is *p*-*d*π bonding.

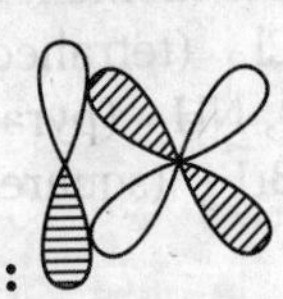

(C) p : ⟶ Both *p* and *d*-orbitals in opposite phase, so it is *p*-*d*π antibonding.

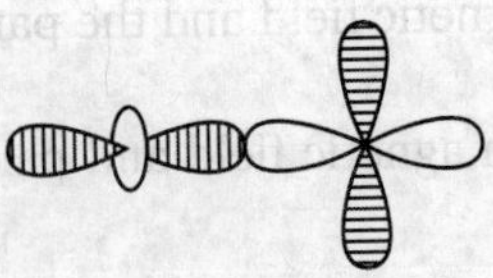

(D) s : ⟶ Both *d*-orbitals axial overlapping in opposite phase. So it is *d*-*d*σ antibonding.

3. (a)

Molecule/Ion	**Hybridisation**	**Shape**
H_3O^+	sp^3	Pyramidal 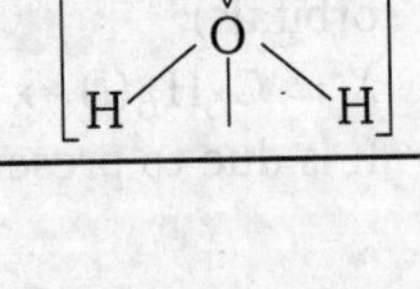

Acetylide	sp	Linear	$\bar{C}\equiv\bar{C}$
NH_4^+	sp^3	Tetrahedral	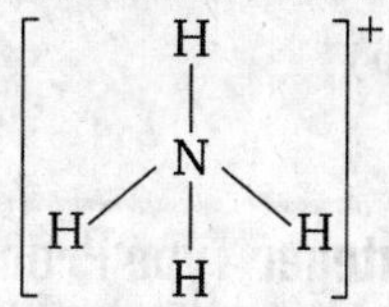
ClO_2^-	sp^3	Bent	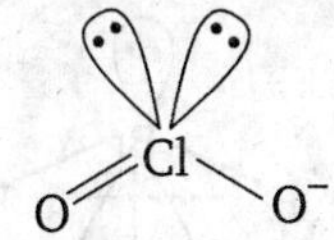

4. (d)

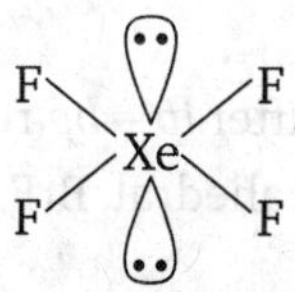

Square planar

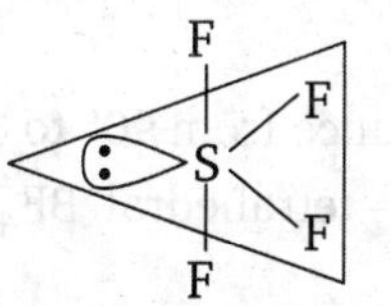

See-saw

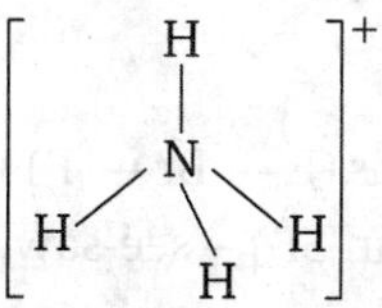

Tetrahedral

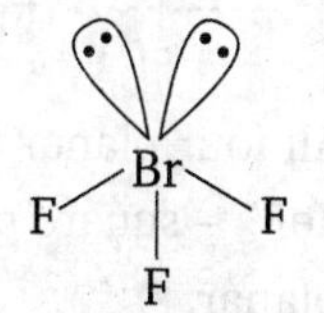

Bent T-shaped

6. (b) (A) $[CoF_6]^{-3}$

$Co^{3+} \rightarrow 3d^6$

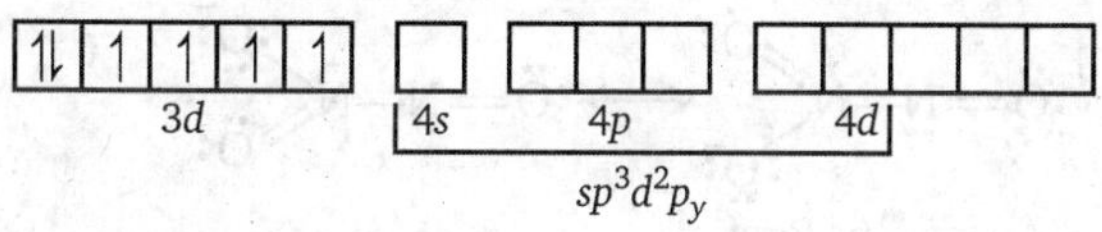

(B) $[NiCl_4]^{2-}$

$Ni^{2+} \rightarrow 3d^8$

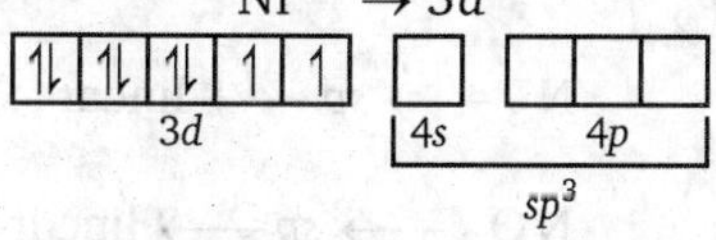

(C) $[Co(NH_3)_6]^{3+}$

$Co^{3+} \rightarrow 3d^6$

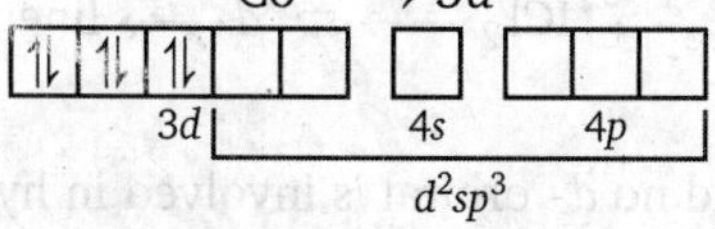

(D) $[Ni(CN)_4]^{2-}$

$Ni^{2+} \rightarrow 3d^8$

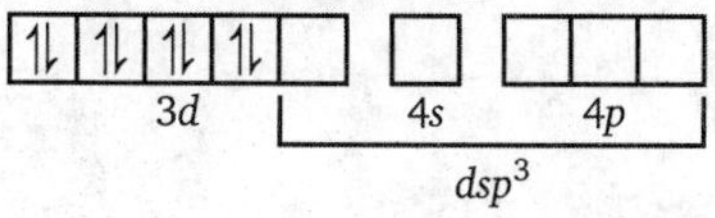

D. Statement and Explanation Type Problems

1. (b)

E. Integer Type Problems

1. (0)

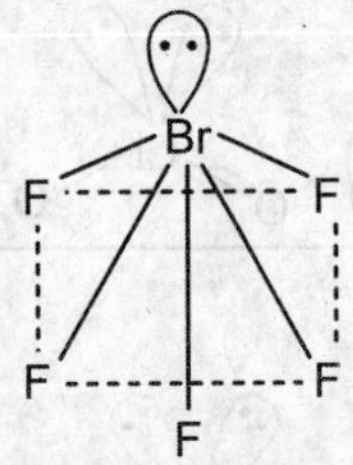

All four planar bonds (F — Br — F) will reduce from 90° to 84.8° after *lp* – *bp* repulsion.

2. (4) XeF_4 – square planar, SF_4 – see-saw, SiF_4 – tetrahedral, BF_4^- – tetrahedral, BrF_4^- – square planar,

$[Cu(NH_3)_4]^{2+}$ – square planar,

$[FeCl_4]^{2-}$ – tetrahedral, $[CoCl_4]^{2-}$ – tetrahedral,

$[PtCl_4]^{2-}$ – square planar.

3. (8)

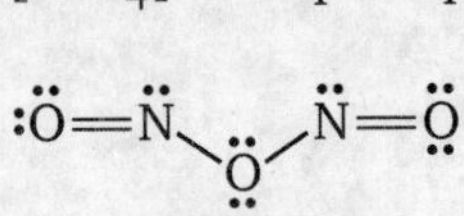

Symmetric form

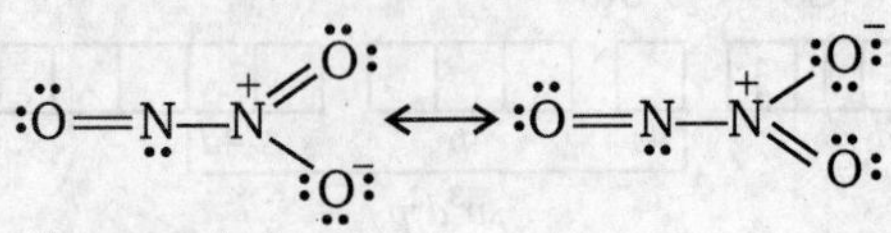

Asymmetric form

Total no. of lone pairs = 8

4. (4) $BeCl_2, N_3^-, N_2O, NO_2^+, O_3, SCl_2, ICl_2^-, I_3^-, XeF_2$

$BeCl_2 \longrightarrow sp \longrightarrow$ linear; $N_3^- \longrightarrow sp \longrightarrow$ linear

$N_2O \longrightarrow sp \longrightarrow$ linear; $\overset{\oplus}{N}O_2 \longrightarrow sp \longrightarrow$ linear

$O_3 \longrightarrow sp^2 \longrightarrow$ bent; $SCl_2 \longrightarrow sp^3 \longrightarrow$ bent

$I_3^- \longrightarrow sp^3d \longrightarrow$ linear; $ICl_2^- \longrightarrow sp^3d \longrightarrow$ linear

$XeF_2 \longrightarrow sp^3d \longrightarrow$ linear

There are four (4) linear molecules/ions and no *d*- orbital is involved in hybridization.

5. (6)

Species		Number of lone pairs
$[TeBr_6]^{2-}$	$\longrightarrow$	1
$[BrF_2]^{1+}$	$\longrightarrow$	2
SNF_3	$\longrightarrow$	0

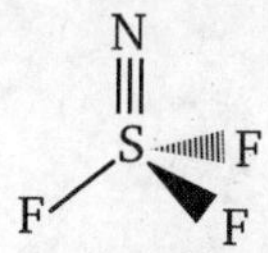 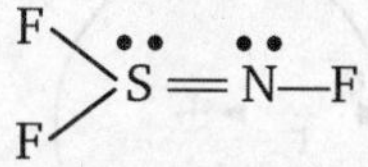

most stable — however an alternate also exists

$[XeF_3]^{1-}$ $\longrightarrow$ 3

$\therefore$ Sum is $= 1 + 2 + 0 + 3 = 6$ lone pair

6. (6) $H_2, Li_2, Be_2, C_2, N_2$ and F_2 are diamagnetic species.

*However because Be_2 does not exists the answer may well be 5.

7. (4) $N \equiv N \rightarrow O$, 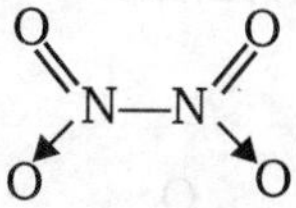

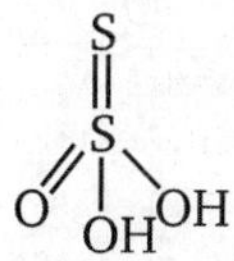 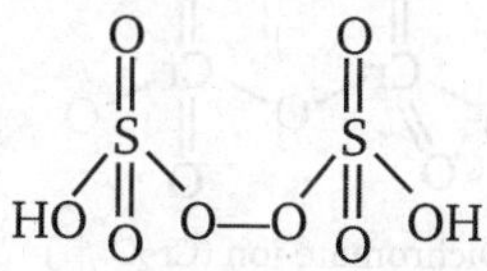

8. (6.75)

$A + B + C \longrightarrow$ Product

$r = k[A]^a [B]^b [C]^c$

By expt. no. 1 and 2, $b = 0$

By expt. no. 1 and 3, $c = 1$

By expt. no. 1 and 4, $a = 1$

$\therefore \quad r = k[A]^1 [B]^0 [C]^1$

From expt. no. 1, $6 \times 10^{-5} = k(0.2)(0.1)$

$\Rightarrow k = 3 \times 10^{-3}$

Now, rate at given concentrations

$$r = k[A]^1 [B]^0 [C]^1 = 3 \times 10^{-3} \times (0.15)^1 \times (0.25)^0 \times (0.15)^1$$

$$= 3 \times 0.0225 \times 10^{-3} = 6.75 \times 10^{-5} \text{ mol dm}^{-3}\text{s}^{-1}$$

$\because \quad Y \times 10^{-5} = 6.75 \times 10^{-5} \text{ mol dm}^{-3}\text{s}^{-1}$

$\therefore \quad Y = 6.75$

9. (6) Here polar molecules in the liquid form will be attracted/deflected near charged comb.

Polar molecules : HF, $H_2O, NH_3, H_2O_2, CHCl_3, C_6H_5Cl$ **(6-polar molecules)**

Non-polar molecules : O_2, CCl_4, C_6H_6

10. (5246.49)

At $d = d_0$, nucleus-nucleus and electron-electron repulsion is absent.

Hence, potential energy will be calculated for 2 H atoms. (P.E. due to attraction of proton and electron),

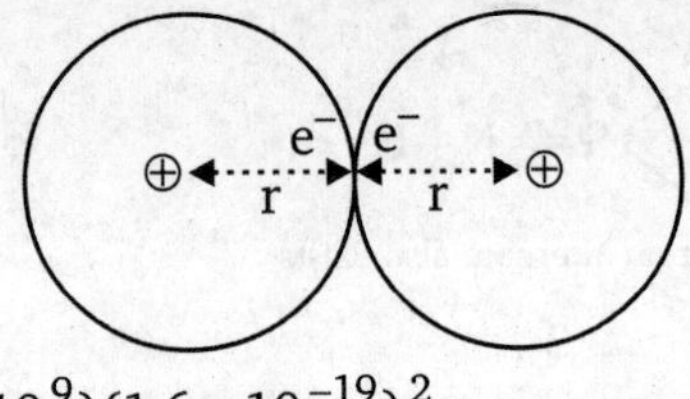

$$\text{P.E.} = \frac{-Kq_1q_2}{\underset{\text{(Bohr radius)}}{r}} = \frac{(9\times10^{9})(1.6\times10^{-19})^2}{0.529\times10^{-10}} = -4.355\times10^{-21}\,\text{kJ}$$

For 1 mol $= -4.355\times10^{-21}\times6.023\times10^{23} = -2623.249\,\text{kJ/mol}$

For 2 H atoms $= -5246.49\,\text{kJ/mol}$

11. (12)

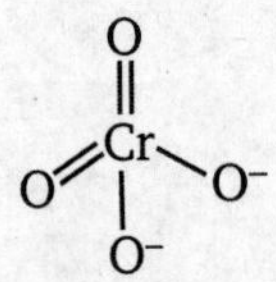

Chromate ion (CrO_4^{2-})

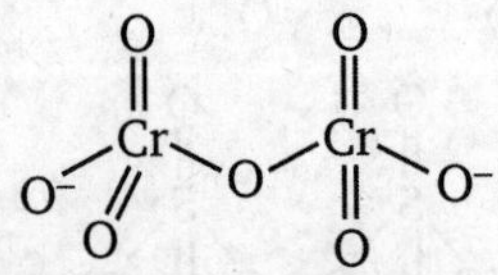

Dichromate ion ($Cr_2O_7^{2-}$)

Total bonds between chromium and oxygen = 4 + 8 = 12

12. (2) SF_4

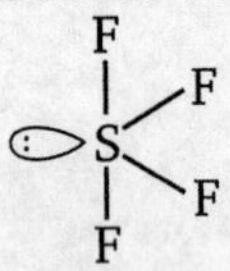

BF_4^-

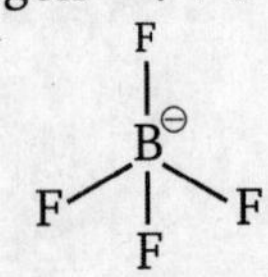

ClF_3

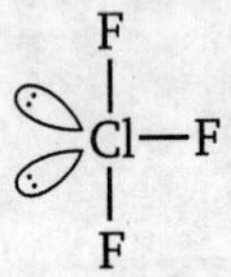

AsF_3

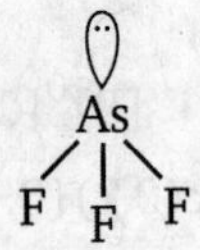

PCl_5

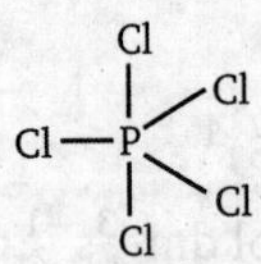

BrF_5

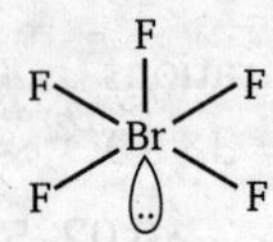

XeF_4

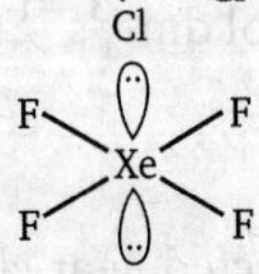

SF_6

13. (19)

$$\underset{(A)}{XeF_6} + 2H_2O \longrightarrow XeO_2F_2 + 4HF$$

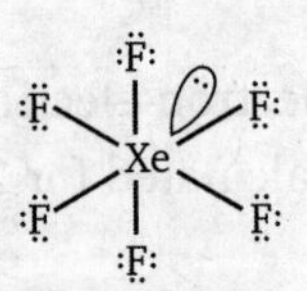

(sp^3d^3)

Distorted octahedral geometry.

Number of lone pairs $= 3\times6+1 = 19$

14. (15)

MOT diagram Bond order of $N_2 = 3$

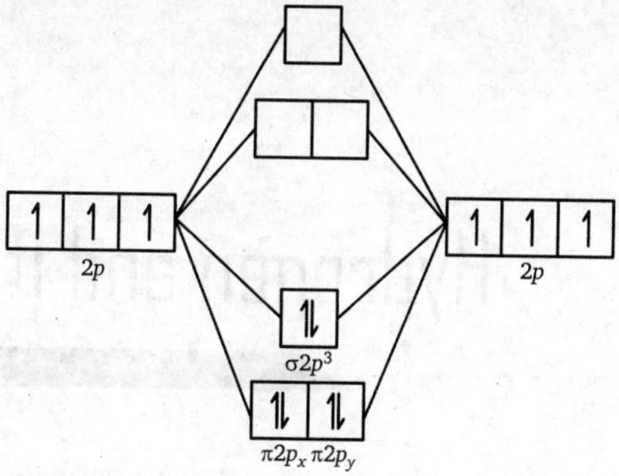

CN will have bond order 2.5 but can not be neutral atom.
NO will also have bond order 2.5 and will exist as neutral atom.
Total electrons in NO is 15.

15. (1) S.N. = 5 [5 B.P. + 0 L. P.]
$PF_5 \Rightarrow sp^3d$ hybridisation

16. (4) H_2O, CO, N_2, NO has two lone pair of electrons.

17. (3)

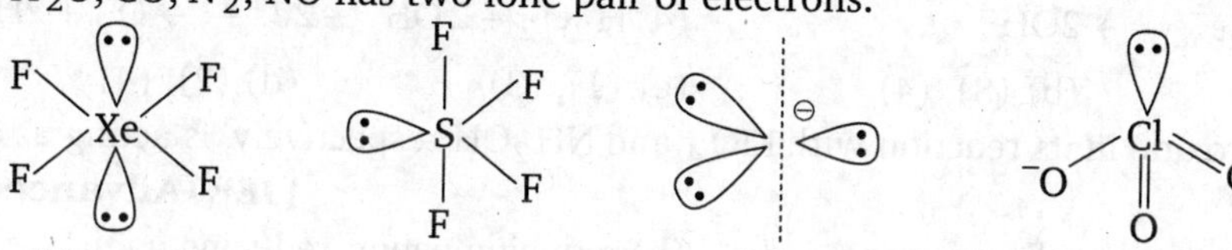

(2 lone pair) (1 lone pair) (3 lone pair) (1 lone pair)

18. (4) Antibonding molecular orbital from $2s = 1$
Antibonding molecular orbital from $2p = 3$
Total = 4

19. (3) PF_5, PCl_5, $Fe(CO)_5$; Trigonal bipyramidal
BrF_5 ; square pyramidal
$[PtCl_4]^{2-}$; square planar
BF_3 ; Trigonal planar

20. (6) CO $\Rightarrow$ $\overset{-}{C}\equiv\overset{+}{O}$: BO = 3
NO^+ $\Rightarrow$ $N\equiv O^+$: BO = 3

21. (6) Linear species are $Cl-Be-Cl$, $O=C=O$, $N^-=N^+=N^-$
(*sp*) (*sp*) (*sp*)

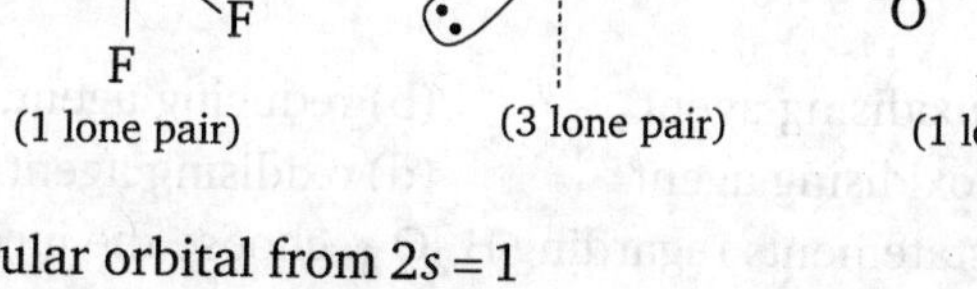

sp^3d sp^3d

❑❑❑

CHAPTER

3

Hydrogen and Its Compounds

A. Objective Questions

[Only one option is correct]

1. In which of the following reactions H_2O_2 acts as a reducing agent? **[JEE (Mains) 2014]**

(1) $H_2O_2 + 2H^+ + 2e^- \longrightarrow 2H_2O$
(2) $H_2O_2 - 2e^- \longrightarrow O_2 + 2H^+$
(3) $H_2O_2 + 2e^- \longrightarrow 2OH^-$
(4) $H_2O_2 + 2OH^- - 2e^- \longrightarrow O_2 + 2H_2O$

(a) (1), (2) (b) (3), (4) (c) (1), (3) (d) (2), (4)

2. Hydrogen peroxide in its reaction with KIO_4 and NH_2OH respectively, is acting as a : **[JEE (Advanced) 2014]**

(a) reducing agent, oxidising agent
(b) reducing agent, reducing agent
(c) oxidising agent, oxidising agent
(d) oxidising agent, reducing agent

3. From the following statements regarding H_2O_2, choose the incorrect statement. **[JEE (Mains) 2015]**

(a) It has to be stored in plastic or wax lined glass bottles in dark.
(b) It has to be kept away from dust.
(c) It can act only as an oxidizing agent.
(d) It decomposes on exposure to light.

4. Which one of the following statements about water is false? **[JEE (Mains) 2016]**

(a) Water is oxidized to oxygen during photosynthesis.
(b) Water can act both as an acid and as a base.
(c) There is extensive intramolecular hydrogen bonding in the condensed phase.
(d) Ice formed by heavy water sinks in normal water.

5. Which of the following combination will produce H_2 gas? **[JEE (Advanced) 2017]**

(a) Cu metal and conc. HNO_3
(b) Zn metal and $NaOH_{(aq)}$
(c) Au metal and $NaCN_{(aq)}$ in the presence of air
(d) Fe metal and conc. HNO_3

6. Hydrogen peroxide oxidises $[Fe(CN)_6]^{4-}$ to $[Fe(CN)_6]^{3-}$ in acidic medium but reduce $[Fe(CN)_6]^{3-}$ to $[Fe(CN)_6]^{4-}$ in alkaline medium. The other products formed are, respectively: **[JEE (Mains) 2018]**

(a) H_2O and $(H_2O + O_2)$
(b) H_2O and $(H_2O + OH^-)$
(c) $(H_2O + O_2)$ and H_2O
(d) $(H_2O + O_2)$ and $(H_2O + OH^-)$

7. The strength of 11.2 volume solution of H_2O_2 is :
[Given that molar mass of $H = 1$ g mol^{-1} and $O = 16$ g mol^{-1}] **[JEE (Mains) 2019]**

(a) 1.7% (b) 13.6% (c) 3.4% (d) 34%

8. The correct statements among (a) to (d) regarding H_2 as a fuel are: **[JEE Main 2019]**

(A) It produces less pollutants than petrol.
(B) A cylinder of compressed dihydrogen weighs ~30 times more than a petrol tank producing the same amount of energy.
(C) Dihydrogen is stored in tanks of metal alloys like $NaNi_5$.
(D) On combustion, values of energy released per gram of liquid dihydrogen and LPG are 50 and 142 kJ, respectively.

(a) (B) and (D) only
(b) (B), (C) and (D) only
(c) (A) and (C) only
(d) (A), (B) and (C) only

9. NaH is an example of: **[JEE Main 2019]**

(a) electron-rich hydride
(b) metallic hydride
(c) molecular hydride
(d) saline hydride

10. The temporary hardness of a water sample is due to compound *X*. Boiling this sample converts *X* to compound *Y*. *X* and *Y*, respectively, are: **[JEE Main 2019]**

(a) $Mg(HCO_3)_2$ and $MgCO_3$
(b) $Ca(HCO_3)_2$ and CaO
(c) $Ca(HCO_3)_2$ and $Ca(OH)_2$
(d) $Mg(HCO_3)_2$ and $Mg(OH)_2$

11. Hydrogen has three isotopes (A), (B) and (C). If the number of neutron(s) in (A), (B) and (C) respectively, are (X), (Y) and (Z), the sum of (X), (Y) and (Z) is: **[JEE Main (Jan.) 2020]**

(a) 4 (b) 3 (c) 2 (d) 1

12. Dihydrogen of high purity (> 99.95%) is obtained through: **[JEE Main (Sept.) 2020]**

(a) the reaction of Zn with dilute HCl.
(b) the electrolysis of acidified water using Pt electrodes
(c) the electrolysis of brine solution.
(d) the electrolysis of warm $Ba(OH)_2$ solution using Ni electrodes.

13. The one that is NOT suitable for the removal of permanent hardness of water is: **[JEE Main (Sept.) 2020]**

(a) Calgon's method
(b) Clark's method
(c) Treatment with sodium carbonate
(d) Ion-exchange method

14. Statements about heavy water are given below. **[JEE Main (Feb.) 2021]**

(A) Heavy water is used in exchange reactions for the study of reaction mechanisms.
(B) Heavy water is prepared by exhaustive electrolysis of water
(C) Heavy water has higher boiling point than ordinary water.
(D) Viscosity of H_2O is greater than D_2O.

Which of the above statements are correct?

(a) A, B and C only
(b) A and B only
(c) A and D only
(d) A and C only

15. The incorrect statement(s) about heavy water is (are): **[JEE Main (March) 2021]**

(A) used as a moderator in nuclear reactor
(B) obtained as a by-product in fertilizer industry.
(C) used for the study of reaction mechanism
(D) has a higher dielectric constant than water

Choose the correct answer from the options given below:

(a) (B) only
(b) (C) only
(c) (D) only
(d) (B) and (D) only

16. Given below are two statements: one is labelled as Assertion A and the other is labelled as Reason R.

Assertion A : Hydrogen is the most abundant element in the Universe, but it is not the most abundant gas in the troposphere.

Reason R : Hydrogen is the lightest element.

In the light of the above statements, choose the correct answer from the options given below:

[JEE Main (Feb.) 2021]

(a) A is true but R is false
(b) Both A and R are true and R is the correct explanation of A
(c) A is false but R is true
(d) Both A and R are true but R is not the correct explanation of A

17. In basic medium, H_2O_2 exhibits which of the following reactions ?

[JEE Main (March) 2021]

(A) $Mn^{2+} \rightarrow Mn^{4+}$ (B) $I_2 \rightarrow I^-$ (C) $PbS \rightarrow PbSO_4$

Choose the most appropriate answer from the options given below :

[JEE Main (March) 2021]

(a) (A), (C) only
(b) (A) only
(c) (B) only
(d) (A), (B) only

18. Which of the following equation depicts the oxidizing nature of H_2O_2 ?

[JEE Main (Feb.) 2021]

(a) $KIO_4 + H_2O_2 \longrightarrow KIO_3 + H_2O + O_2$
(b) $2I^- + H_2O_2 + 2H^+ \longrightarrow I_2 + 2H_2O$
(c) $I_2 + H_2O_2 + 2OH^- \longrightarrow 2I^- + 2H_2O + O_2$
(d $Cl_2 + H_2O_2 \longrightarrow 2HCl + O_2$

19. Given below are two statements :

Statement I : H_2O_2 can act as both oxidising and reducing agent in basic medium.

Statement II : In the hydrogen economy, the energy is transmitted in the form of dihydrogen.

In the light of the above statements, choose the correct answer from the options given below :

[JEE Main (March) 2021]

(a) Both statement I and statement II are false.
(b) Both statement I and statement II are true.

(c) Statement I is true but statement II is false.
(d) Statement I is false but statement II is true.

20. The correct statements about H_2O_2 are :
(A) used in the treatment of effluents.
(B) used as both oxidising and reducing agents.
(C) the two hydroxyl groups lie in the same plane.
(D) miscible with water.
Choose the correct answer from the options given below : **[JEE Main (March) 2021]**
(a) (A), (B), (C) and (D) (b) (A), (B) and (D) only
(c) (B), (C) and (D) only (d) (A), (C) and (D) only

21. The highest industrial consumption of molecular hydrogen is to produce compounds of element : **[JEE Main (June) 2022 (I)]**
(a) Carbon (b) Nitrogen (c) Oxygen (d) Chlorine

22. High purity (>99.95%) dihydrogen is obtained by : **[JEE Main (July) 2022 (II); (Sep.) 2022]**
(a) reaction of zinc with aqueous alkali.
(b) electrolysis of acidified water using platinum electrodes.
(c) electrolysis of warm aqueous barium hydroxide solution between nickel electrodes.
(d) reaction of zinc with dilute acid.

23. Consider the following reaction :

$$2HSO_4^-(aq) \xrightarrow[\text{(2) Hydrolysis}]{\text{(1) Electrolysis}} 2HSO_4^- + 2H^+ + A$$

The dihedral angle in product A in its solid phase at 110 K is : **[JEE Main (June) 2022 (I)]**
(a) 104° (b) 111.5° (c) 90.2° (d) 111.0°

24. The products obtained from a reaction of hydrogen peroxide and acidified potassium permanganate are : **[JEE Main (July) 2022 (II)]**
(a) Mn^{4+}, H_2O only (b) Mn^{2+}, H_2O only
(c) Mn^{4+}, H_2O, O_2 only (d) Mn^{2+}, H_2O, O_2 only

25. The reaction of H_2O_2 with potassium permanganate in acidic medium leads to the formation of mainly : **[JEE Main (July) 2022 (I)]**
(a) Mn^{2+} (b) Mn^{4+} (c) Mn^{3+} (d) Mn^{6+}

26. Given below are two reactions, involved in the commercial production of dihydrogen (H_2). The two reactions are carried out at temperature "T_1" and "T_2" respectively.

$$C(s) + H_2O(g) \xrightarrow{T_1} CO(g) + H_2(g)$$

$$CO(g) + H_2O(g) \xrightarrow[\text{Catalyst}]{T_2} CO_2(g) + H_2(g)$$

The temperature T_1 and T_2 are correctly related as : **[JEE Main (April) 2023 (I)]**
(a) $T_1 > T_2$ (b) $T_1 = T_2$
(c) $T_1 = 100$ K, $T_2 = 1270$ K (d) $T_1 < T_2$

27. The water gas on reacting with cobalt as a catalyst forms : **[JEE Main (April) 2023 (I)]**

(a) Ethanol (b) Methanoic acid

(c) Methanal (d) Methanol

28. The starting material for convenient preparation of deuterated hydrogen peroxide (D_2O_2) in laboratory is : **[JEE Main (Feb.) 2023 (II)]**

(a) $K_2S_2O_8$ (b) BaO_2

(c) BaO (d) 2-ethylanthraquinol

29. '25 volume' hydrogen peroxide means : **[JEE Main (Jan.) 2023 (I)]**

(a) 1 L marketed solution contains 250 g of H_2O_2.

(b) 1 L marketed solution contains 75 g of H_2O_2.

(c) 100 mL marketed solution contains 25 g of H_2O_2.

(d) 1 L marketed solution contains 25 g of H_2O_2.

30. Choose the correct statements about the hydrides of group 15 elements.

A. The stability of the hydrides decreases in the order $NH_3 > PH_3 > AsH_3 > SbH_3 > BiH_3$

B. The reducing ability of the hydrides increases in the order $NH_3 < PH_3 < AsH_3 < SbH_3 < BiH_3$

C. Among the hydrides, NH_3 is strong reducing agent while BiH_3 is mild reducing agent

D. The basicity of the hydrides increases in the order $NH_3 < PH_3 < AsH_3 < SbH_3 < BiH_3$

Choose the most appropriate from the option given below : **[JEE Main (Jan.) 2024]**

(a) B and C only (b) C and D only

(c) A and B only (d) A and D only

31. The strongest reducing agent among the following is : **[JEE Main (Jan.) 2024]**

(a) NH_3 (b) SbH_3

(c) BiH_3 (d) PH_3

B. Objective Questions

[One or more than one option(s) is/are correct]

1. Hydrogen bonding plays a central role in the following phenomena : **[JEE (Advanced) 2014]**

(a) Ice floats in water.

(b) Higher Lewis basicity of primary amines than tertiary amines in aqueous solutions.

(c) Formic acid is more acidic than acetic acid.

(d) Dimerisation of acetic acid in benzene.

C. Matching Type Problems

1. All the compounds listed in Column I react with water. Match the result of the respective reactions with the appropriate options listed in Column II. **(IIT 2010)**

	Column I		Column II
(a)	$(CH_3)_2SiCl_2$	**(p)**	Hydrogen halide formation
(b)	XeF_4	**(q)**	Redox reaction
(c)	Cl_2	**(r)**	Reacts with glass
(d)	VCl_5	**(s)**	Polymerization
		(t)	O_2 formation

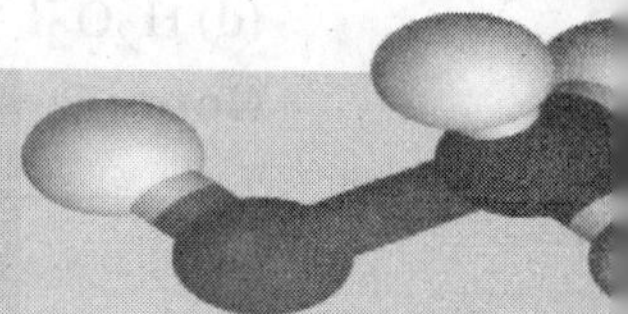

Answers

[A] Objective Questions (Only one option is correct)

1. (d) **2.** (a) **3.** (c) **4.** (c) **5.** (b) **6.** (a) **7.** (c) **8.** (d) **9.** (d) **10.** (d)
11. (b) **12.** (d) **13.** (a) **14.** (a) **15.** (c) **16.** (b) **17.** (d) **18.** (b) **19.** (b) **20.** (b)
21. (b) **22.** (c) **23.** (c) **24.** (d) **25.** (a) **26.** (a) **27.** (d) **28.** (a) **29.** (b) **30.** (c)
31. (c)

[B] Objective Questions (One or more than one option(s) is/are correct)

1. (a, b, d)

[C] Matching Type Problems

1. (a) $\rightarrow$ p, s; (b) $\rightarrow$ p, q, r, t; (c) $\rightarrow$ p, q, t; (d) $\rightarrow$ p

Hints and Solutions

A. Objective Questions

[Only one option is correct]

1. (d) The reducing agent oxidises itself.

(a) $H_2O_2^{-1} + 2H^+ + 2e^- \longrightarrow 2H_2O^{2-}$

(b) $H_2O_2^{-1} - 2e \longrightarrow O_2^0 + 2H^+$

(c) $H_2O_2^{-1} + 2e \longrightarrow 2\overset{2-}{O}H^-$

(d) $H_2O_2^{-1} + 2OH^- - 2e \longrightarrow O_2^0 + H_2O$

Note : Powers of 'O' are oxidation number of 'O' in the compound.

Decrease in O.N.

2. (a) $\overset{+7}{KIO_4} + \underset{\text{(Reducing agent)}}{H_2O_2} \longrightarrow \overset{+5}{KIO_3} + H_2O + O_2$

$\overset{-1}{NH_2OH} + \underset{\text{(Oxidising agent)}}{3H_2O_2} \longrightarrow \overset{+5}{HNO_3} + 4H_2O$

Increase in O.N.

3. (c) H_2O_2 acts as an oxidising as well as a reducing agent.

4. (c) In the condensed phase, there is extensive intermolecular hydrogen bonding in water molecules but not intramolecular hydrogen bonding.

5. (b) $Zn + 2NaOH \longrightarrow Na_2ZnO_2 + H_2\uparrow$

6. (a) During reduction $H_2O_2 \longrightarrow H_2O$

During reduction $H_2O_2 \longrightarrow O_2$

7. (c) $\underset{\substack{2 \times 34g \\ =68g \text{ at STP}}}{2H_2O_{2(l)}} \longrightarrow \underset{22.4L}{O_{2(g)}} + H_2O_{(l)}$

22.4 L of O_2 at STP produced from 68 g of H_2O_2

11.2 L of O_2 at STP product from $= \dfrac{68 \times 11.2}{22.4} = 34$ g of H_2O_2

Therefore, strength of H_2O_2 in 11.2 volume H_2O_2 solution $= 34 g/L \equiv 3.4\%$ H_2O_2 solution.

8. (d) $H_2(g) + \frac{1}{2}O_2(g) \longrightarrow H_2O(l)$ less pollution.

It is stored in $NaNi_5$, $Ti - TiH_2$, $Mg - MgH_2$ metal alloys.

9. (d) Saline hydride are also known as ionic hydride and we know that NaH is an example of ionic hydride.

10. (d) Temporary hardness is due to presence of magnesium and calcium hydrogen carbonates

$$Mg(HCO_3)_2 \xrightarrow{\Delta} Mg(OH)_2 \downarrow + 2CO_2 \uparrow$$

$$Ca(HCO_3)_2 \xrightarrow{\Delta} CaCO_3 \downarrow + H_2O + CO_2 \uparrow$$

11. (b) Hydrogen have 3 isotopes:

Isotopes	Neutrons
${}^1_1H$	$x = 0$
${}^2_1H$	$y = 1$
${}^3_1H$	$z = 2$

Total neutrons $= 0 + 1 + 2 = 3$

12. (d) Dihydrogen of high purity (> 99.95%) is obtained through the electrolysis of warm $Ba(OH)_2$ solution using Ni electrodes.

13. (a) Clark's method is not suitable for removal of permanent hardness as this method is applied for temporary hardness.

14. (a) Viscosity of D_2O is greater than that of H_2O due to its high molecular mass.

15. (c) Dielectric constant of H_2O is greater than that of D_2O due to less polarisation in D_2O.

16. (b) Most abundant element in the universe is hydrogen, where as most abundant gas in troposphere is nitrogen not hydrogen.

- Hydrogen is lightest element that comes first in periodic table with atomic mass equals to 1.008 amu.

17. (d) Oxidising action of H_2O_2 in basic medium:

$$Mn^{2+} + H_2O_2 \longrightarrow Mn^{4+} + 2OH^-$$

Reducing action of H_2O_2 in basic medium :

$$I_2 + H_2O_2 + 2OH^- \longrightarrow 2I^- + 2H_2O + O_2$$

In acidic medium, H_2O_2 oxidises PbS to $PbSO_4$:

$$PbS + 4H_2O_2 \longrightarrow PbSO_4 + 4H_2O$$

18. (b)

$$\overset{+7}{KIO_4} + \underset{(R.A.)}{H_2O_2} \longrightarrow \overset{+5}{KIO_3} + H_2O + O_2$$

$$\overset{-1}{2I^-} + \underset{(O.A.)}{H_2O_2} + 2H^+ \longrightarrow \overset{0}{I_2} + 2H_2O$$

$$\overset{0}{I_2} + \underset{(R.A.)}{H_2O_2} + 2OH^- \longrightarrow \overset{-1}{2I} + 2H_2O + O_2$$

$$\overset{0}{Cl_2} + \underset{\text{(R.A.)}}{\overset{-1}{H_2O_2}} \longrightarrow 2HCl + O_2$$

19. (b) Statement I and statement II are true. H_2O_2 acts as both oxidising and reducing agent since in H_2O_2, oxygen atom is present in intermediate oxidation state −1.
Hydrogen economy is related with transportation and storage of hydrogen as a fuel.

20. (b) H_2O_2 used in treatment of effluent due to its oxidising nature. Since, oxygen atom is present in −1 oxidation state, it acts as both oxidising and reducing agent. H_2O_2 molecule is non-planar so, two hydroxyl groups lie in different plane.

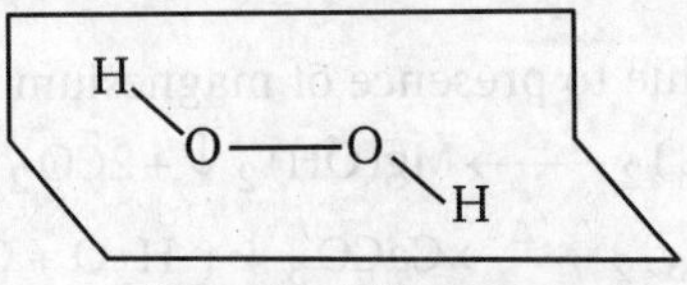

(Open book like structure)

H_2O_2 is miscible in water due to formation of hydrogen bonding.

21. (b) Around 55% of hydrogen consumed in ammonia production.

22. (c) Electrolysis of warm aqueous $Ba(OH)_2$ solution using nickel electrode yields high purity (>99.95%) dihydrogen.

23. (c) $$2HSO_4^-(aq.) \xrightarrow[\text{(2) Hydrolysis}]{\text{(1) Electrolysis}} 2HSO_4^- + 2H^+ + \underset{(A)}{H_2O_2}$$

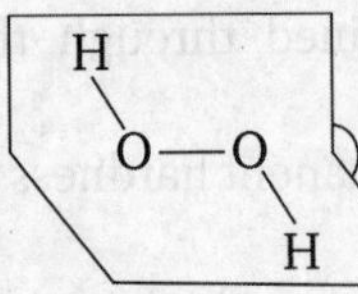

90.2° solid phase.

24. (d) $6H^+ + 2MnO_4^- + 5H_2O_2 \longrightarrow 2Mn^{2+} + 8H_2O + 5O_2$

25. (a) In acidic medium MnO_4^- forms Mn^{2+} by reacting with H_2O_2.

26. (a) $T_1 = 1270$ K, $T_2 = 673$ K

$T_1 > T_2$

27. (d) $CO + 2H_2 \xrightarrow{Co} CH_3OH$

28. (a) $K_2S_2O_8(s) + 2D_2O(l) \longrightarrow 2KDSO_4(aq.) + D_2O_2$

29. (b) '25 volume' H_2O_2 means 1 L of H_2O_2 solution produces 25 L of O_2.

$$\underset{\substack{2 \times 34 \\ = 68\text{ g}}}{2H_2O_2} \longrightarrow 2H_2O + \underset{\substack{22.7\text{ L} \\ \text{(at STP)}}}{O_2}$$

$\therefore$ 25 L of O_2 produced by $= \frac{68 \times 25}{22.7}$ g $= 74.89 = 75$ g of H_2O_2

30. (c) On moving down the group, bond strength of M—H bond decreases, which reduces the thermal stability but increases reducing nature of hydrides, hence *A* and *B* are correct statements.

31. (c) Strongest reducing agent : BiH_3 explained by its low bond dissociation energy.

B. Objective Questions

[One or more than one option(s) is/are correct]

1. (a, b, d)

⇒ Ice floats in water due to the low density of ice as compare to water which is due to open cage like structuer (formed by intermolecular H-bonding).

⇒ Dimerisation of acetic acid in benzene is due to intermolecular hydrogen bonding.

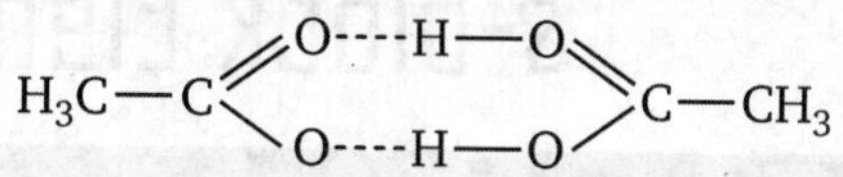

⇒ Basic strength of $RNH_2 > R_3N$ it also explained by hydrogen bonding.

C. Matching Type Problems

1. (a)—p, s; (b)—p, q, r, t; (c)—p,q,t; (d)—p

(a) $(CH_3)_2SiCl_2 + 2H_2O \longrightarrow (CH_3)_2Si(OH)_2 + 2HCl$

$(CH_3)_2Si(OH)_2$ can undergo polymerisation to form silicones.

(b) $3XeF_4 + 6H_2O \longrightarrow XeO_3 + 2Xe + 12HF + 1\frac{1}{2}O_2$

(c) $Cl_2 + H_2O \longrightarrow 2HCl + \frac{1}{2}O_2$

(d) $VCl_5 + H_2O \longrightarrow COCl_3 + 2HCl$

❑❑❑

CHAPTER

4

s-block Elements

A. Objective Questions

[Only one option is correct]

1. Which of the following represents the correct order of increasing first ionization enthalpy for Ca, Ba, S, Se and Ar? **[JEE (Mains) 2013]**

(a) Ca < S < Ba < Se < Ar
(b) S < Se < Ca < Ba < Ar
(c) Ba < Ca < Se < S < Ar
(d) Ca < Ba < S < Se < Ar

2. Which one of the following alkaline earth metal sulphates has its hydration enthalpy greater than its lattice enthalpy? **[JEE (Mains) 2015]**

(a) $BaSO_4$
(b) $SrSO_4$
(c) $CaSO_4$
(d) $BeSO_4$

3. The main oxides formed on combustion of Li, Na and K in excess of air are, respectively : **[JEE (Mains) 2016]**

(a) Li_2O, Na_2O and KO_2
(b) LiO_2, Na_2O_2 and K_2O
(c) Li_2O_2, Na_2O_2 and KO_2
(d) Li_2O, Na_2O_2 and KO_2

4. The correct order of hydration enthalpies of alkali metal ions is : **[JEE (Mains) 2019]**

(a) $Li^+ > Na^+ > K^+ > Rb^+ > Cs^+$
(b) $Na^+ > Li^+ > K^+ > Rb^+ > Cs^+$
(c) $Na^+ > Li^+ > K^+ > Cs^+ > Rb^+$
(d) $Li^+ > Na^+ > K^+ > Cs^+ > Rb^+$

5. The covalent alkaline earth metal halide (X = Cl, Br, I) is : **[JEE (Mains) 2019]**

(a) CaX_2
(b) BeX_2
(c) MgX_2
(d) SrX_2

6. On combustion of Li, Na and K in excess of air, the major oxides formed, respectively, are: **[JEE Main (Sept.) 2020]**

(a) Li_2O, Na_2O_2 and K_2O
(b) Li_2O, Na_2O_2 and KO_2
(c) Li_2O, Na_2O and K_2O_2
(d) Li_2O, Na_2O_2 and K_2O_2

7. Among the statements (A)–(D), the correct ones are: **[JEE Main (Jan.) 2020]**

(A) Lithium has the highest hydration enthalpy among the alkali metals.
(B) Lithium chloride is insoluble in pyridine.
(C) Lithium cannot form ethynide upon its reaction with ethyne
(D) Both lithium and magnesium react slowly with H_2O

(a) (A), (B) and (D) only
(b) (A) and (D) only
(c) (B) and (C) only
(d) (A), (C) and (D) only

8. Two elements *A* and *B* have similar chemical properties. They don't form solid hydrogencarbonates, but react with nitrogen to form nitrides. *A* and *B*, respectively, are:
[JEE Main (Sept.) 2020]

(a) Na and Ca
(b) Na and Rb
(c) Cs and Ba
(d) Li and Mg

9. A metal (*A*) on heating in nitrogen gas gives compound *B*. *B* on treatment with H_2O gives a colourless gas which when passed through $CuSO_4$ solution gives a dark blue-violet coloured solution. *A* and *B* respectively, are:
[JEE Main (Jan.) 2020]

(a) Mg and $Mg(NO_3)_2$
(b) Na and $NaNO_3$
(c) Mg and Mg_3N_2
(d) Na and Na_3N

10. An alkaline earth metal '*M*' readily forms water soluble sulphate and water insoluble hydroxide. Its oxide *MO* is very stable to heat and does not have rock-salt structure. *M* is :
[JEE Main (Sept.) 2020]

(a) Sr
(b) Be
(c) Mg
(d) Ca

11. Match the following compounds (Column-I) with their uses (Column-II):
[JEE Main (Sept.) 2020]

	Column-I		Column-II
(i)	$Ca(OH)_2$	(A)	Casts of statues
(ii)	NaCl	(B)	White wash
(iii)	$CaSO_4 \cdot \frac{1}{2}H_2O$	(C)	Antacid
(iv)	$CaCO_3$	(D)	Washing soda preparation

(a) (i) → (D); (ii) → (A); (iii) → (C); (iv) → (B)
(b) (i) → (B); (ii) → (D); (iii) → (A); (iv) → (C)
(c) (i) → (B); (ii) → (C); (iii) → (D); (iv) → (A)
(d) (i) → (C); (ii) → (D); (iii) → (B); (iv) → (A)

12. When gypsum is heated to 393 K, it forms:
[JEE Main (Jan.) 2020]

(a) $CaSO_4 \cdot 0.5H_2O$
(b) Dead burnt plaster
(c) Anhydrous $CaSO_4$
(d) $CaSO_4 \cdot 5H_2O$

13. Match List-I with List-II :

	List-I (Salt)		List-II (Flame colour wavelength)
(A)	LiCl	(i)	455.5 nm
(B)	NaCl	(ii)	670.8 nm
(C)	RbCl	(iii)	780.0 nm
(D)	CsCl	(iv)	589.2 nm

Choose the correct answer from the options given below : **[JEE Main (Feb.) 2021]**

(a) (A) → (iv), (B) → (ii), (C) → (iii), (D) → (i)
(b) (A) → (ii), (B) → (i), (C) → (iv), (D) → (iii)
(c) (A) → (i), (B) → (iv), (C) → (ii), (D) → (iii)
(d) (A) → (ii), (B) → (iv), (C) → (iii), (D) → (i)

14. Given below are two statements :

Statement I : Both $CaCl_2 \cdot 6H_2O$ and $MgCl_2 \cdot 8H_2O$ undergo dehydration on heating.

Statement II : BeO is amphoteric whereas the oxides of other elements in the same group are acidic.

In the light of the above statements, choose the correct answer from the options given below : **[JEE Main (March) 2021]**

(a) Statement I is false but statement II is true.
(b) Both statement I and statement II are false.
(c) Both statement I and statement II are true.
(d) Statement I is true but statement II is false.

15. Given below are two statements : One is labelled as Assertion A and the other labelled as reason R.

Assertion A: During the boiling of water having temporary hardness, $Mg(HCO_3)_2$ is converted to $MgCO_3$.

Reason R: The solubility product of $Mg(OH)_2$ is greater than that of $MgCO_3$.

In the light of the above statements, choose the most appropriate answer from the options given below: **[JEE Main (March) 2021]**

(a) Both A and R are true but R is not the correct explanation of A.
(b) A is true but R is false.
(c) Both A and R are true and R is the correct explanation of A.
(d) A is false but R is true.

16. Match List-I with List-II. **[JEE Main (March) 2021]**

	List-I		List-II
(A)	$Ca(OCl)_2$	(i)	Antacid
(B)	$CaSO_4 \cdot \frac{1}{2}H_2O$	(ii)	Cement
(C)	CaO	(iii)	Bleach
(D)	$CaCO_3$	(iv)	Plaster of Paris

Choose the most appropriate answer from the options given below:

(a) A-i, B-iv, C-iii, D-ii
(b) A-iii, B-ii, C-iv, D-i
(c) A-iii, B-iv, C-ii, D-i
(d) A-iii, B-ii, C-i, D-iv

17. One of the by-products formed during the recovery of NH_3 from solvay process is: **[JEE Main (March) 2021]**

(a) $Ca(OH)_2$ (b) $NaHCO_3$ (c) $CaCl_2$ (d) NH_4Cl

18. Lithium nitrate and sodium nitrate, when heated separately, respectively, give : **[JEE Main (July) 2022 (I)]**

(a) $LiNO_2$ and $NaNO_2$ (b) Li_2O and Na_2O
(c) Li_2O and $NaNO_2$ (d) $LiNO_2$ and Na_2O

19. Correct statement about alkali metal oxides is : **[JEE Main (June) 2022 (I)]**

(a) peroxides are coloured
(b) superoxides are paramagnetic
(c) oxides are paramagnetic
(d) peroxides are both coloured and paramagnetic

20. Portland cement contains 'X' to enhance the setting time. What is 'X'? **[JEE Main (July) 2022 (II)]**

(a) $CaSO_4 \cdot \frac{1}{2} H_2O$ (b) $CaSO_4 \cdot 2H_2O$ (c) $CaSO_4$ (d) $CaCO_3$

21. Reaction of $BeCl_2$ with $LiAlH_4$ gives : **[JEE Main (July) 2022 (I)]**

(A) $AlCl_3$ (B) BeH_2 (C) LiH (D) LiCl
(E) $BeAlH_4$

Choose the correct answer from options given below :

(a) (A), (D) and E (b) (A), (B) and (D) (c) (D) and (E) (d) (B), (C) and (D)

22. Addition of H_2SO_4 to BaO_2 produces: **[JEE Main (June) 2022 (I)]**

(a) BaO, SO_2 and H_2O (b) $BaHSO_4$ and O_2
(c) $BaSO_4$, H_2 and O_2 (d) $BaSO_4$ and H_2O_2

23. Identify the correct order of standard enthalpy of formation of sodium halides. **[JEE Main (April) 2023 (II)]**

(a) NaI < NaBr < NaCl < NaF (b) NaF < NaCl < NaBr < NaI
(c) NaCl < NaF < NaBr < NaI (d) NaI < NaBr < NaF < NaCl

24. Alkali metal from the following with least melting point is : **[JEE Main (April) 2023 (II)]**

(a) Rb (b) K (c) Na (d) Cs

25. Better method for preparation of BeF_2, among the following is : **[JEE Main (April) 2023 (II)]**

(a) $(NH_4)_2BeF_4 \xrightarrow{\Delta} BeF_2$ (b) $BeH_2 + F_2 \xrightarrow{\Delta} BeF_2$

(c) $Be + F_2 \xrightarrow{\Delta} BeF_2$ (d) $BeO + C + F_2 \xrightarrow{\Delta} BeF_2$

26. The compound which does not exist is : **[JEE Main (April) 2023 (I)]**

(a) NaO_2 (b) $(NH_4)_2BeF_4$ (c) BeH_2 (d) $PbEt_4$

27. Structure of $BeCl_2$ in solid state, vapour phase and at very high temperature respectively are : **[JEE Main (April) 2023 (II)]**

(a) Dimeric, Polymeric, Monomeric (b) Polymeric, Dimeric, Monomeric
(c) Monomeric, Dimeric, Polymeric (d) Polymeric, Monomeric, Dimeric

28. Given below are two statements:

Statement I : The 4*f* and 5*f*-series of elements are placed separately in the periodic table to preserve the principle of classification.

Statement II : *s*-block elements can be found in pure form in nature. In the light of the above statements, choose the most appropriate answer from the options given below :

[JEE Main (Jan.) 2024]

(a) Statement I is false but Statement II is true
(b) Both Statement I and Statement II are true
(c) Statement I is true but Statement II is false
(d) Both Statement I and Statement II are false

B. Objective Questions

[One or more than one option(s) is/are correct]

1. The compound(s) formed upon combustion of sodium metal in excess air is(are): **(IIT 2009)**
(a) Na_2O_2 (b) Na_2O (c) NaO_2 (d) NaOH

C. Statement and Explanation Type Problems

Read the following questions and answer as per the direction given below :
(a) Statement 1 is true; Statement 2 is true; Statement 2 is a correct explanation of Statement 1.
(b) Statement 1 is true; Statement 2 is true; Statement 2 is not the correct explanation of Statement 1.
(c) Statement 1 is true; Statement 2 is false.
(d) Statement 1 is false; Statement 2 is true.

1. **Statement 1 :** Alkali metals dissolve in liquid ammonia to give blue solutions.
Statement 2 : Alkali metals in liquid ammonia give solvated species of the type $[M(NH_3)_n]^+$ (M = alkali metals). **(IIT 2007)**

Answers

[A] Objective Questions (Only one option is correct)

1. (c)	**2.** (d)	**3.** (d)	**4.** (a)	**5.** (b)	**6.** (b)	**7.** (d)	**8.** (d)	**9.** (c)	**10.** (b)
11. (b)	**12.** (c)	**13.** (d)	**14.** (b)	**15.** (d)	**16.** (c)	**17.** (c)	**18.** (c)	**19.** (b)	**20.** (b)
21. (b)	**22.** (d)	**23.** (a)	**24.** (d)	**25.** (a)	**26.** (a)	**27.** (b)	**28.** (c)		

[B] Objective Questions (One or more than one option(s) is/are correct)

1. (a, b)

[C] Statement and Explanation Type Problems

1. (a)

Hints and Solutions

A. Objective Questions

[Only one option is correct]

1. (c) Factual.

2. (d) Be^{2+} being smaller in size has maximum hydration enthalpy which exceeds its lattice enthalpy.

3. (d)

$$Li + O_2 \longrightarrow \underset{\text{Normal oxide}}{Li_2O}$$

$$2Na + O_2 \longrightarrow \underset{\text{Peroxide}}{Na_2O_2}$$

$$K + O_2 \longrightarrow \underset{\text{Superoxide}}{KO_2}$$

4. (a) As the degree of hydration decreases from Li^+ to Cs^+, the hydration energy of alkali metal ions also decreases from Li^+ to Cs^+.

5. (b) Except for beryllium (Be) halides, all other halides of alkaline earth metals are ionic in nature.

6. (b) Major oxides formed are:

$$2Li + \frac{1}{2}O_2 \longrightarrow Li_2O$$

$$2Na + O_2 \longrightarrow Na_2O_2$$

$$K + O_2 \longrightarrow KO_2$$

7. (d) (A) Lithium due to small size of Li^+ has highest hydration enthalpy.

(B) LiCl is soluble in organic solvent like pyridine.

(C) Lithium is only alkali metal which does not react with ethyne to form ethynide.

(D) Li and Mg show diagonal relationship and react slowly with H_2O.

8. (d) Li and Mg show similar chemical properties due to diagonal relationship also they don't form solid hydrogen carbonates, Li_3N and Mg_3N_2 are their nitride.

9. (c)

$$\underset{(A)}{3Mg} + N_2 \longrightarrow \underset{(B)}{Mg_3N_2}$$

$$\underset{(B)}{Mg_3N_2} + 6H_2O \longrightarrow 3Mg(OH)_2 + \underset{\text{Colourless gas}}{2NH_3}$$

$$CuSO_4 + 4NH_3 \longrightarrow \underset{\text{Blue-violet colour solution}}{[Cu(NH_3)_4]SO_4}$$

10. (b) BeO is stable to heat and have wurtzite type structure.

11. (b) Correct combination is

$$Ca(OH)_2 \longrightarrow \text{white wash}$$

$$NaCl \longrightarrow \text{washing soda preparation}$$

$$CaSO_4 \cdot \frac{1}{2}H_2O \longrightarrow \text{casts of statues}$$

$$CaCO_3 \longrightarrow \text{antacid.}$$

12. (c) When gypsum is heated above 393 K, anhydrous $CaSO_4$ is formed.

$$CaSO_4 \cdot 2H_2O(s) \xrightarrow{T>393K} CaSO_4(s) + 2H_2O(l)$$

13. (d)

Element	Flame colour	Wavelength (nm)
Li	Crimson red	670.8
Na	Yellow	589.2
Rb	Red violet	780.0
Cs	Blue	455.5

14. (b) Statement I is false only $CaCl_2 \cdot 6H_2O$ undergoes dehydration on hydrolysis.

$$CaCl_2 \cdot 6H_2O(s) \xrightarrow{\Delta} CaCl_2(s) + 6H_2O(l)$$

Statement II is also incorrect since other elements of group (II) oxides are basic in nature. (Oxides of metals are basic in nature).

15. (d) $Mg(HCO_3)_2 \xrightarrow{\Delta} Mg(OH)_2 \downarrow + 2CO_{2(g)}$

So, Assertion is false.

Since, $K_{sp}Mg(OH)_2 > K_{sp}MgCO_3$

So, that $Mg(OH)_2$ precipitate first. Reason is true.

16. (c) $CaOCl_2 \longrightarrow$ Bleaching powder

$CaSO_4 \cdot \frac{1}{2}H_2O \longrightarrow$ Plaster of Paris

$CaO \longrightarrow$ Used in cement manufacturing

$CaCO_3 \longrightarrow$ Antacid

17. (c) Solvay process; $Ca(OH)_2 + 2NH_4Cl \longrightarrow CaCl_2 + 2NH_3 + 2H_2O$

18. (c) $4LiNO_3 \longrightarrow 2Li_2O + 4NO_2 + O_2$

$2NaNO_3 \longrightarrow 2NaNO_2 + O_2$

19. (b) The peroxide and oxides of alkali metals are colourless when pure.

Superoxides are paramagnetic while peroxides are diamagnetic.

Electronic configuration of O_2^{2-} (peroxide)

$(\sigma 1s^2)(\sigma * 1s^2)(\sigma 2s^2)(\sigma * 2s^2)(\sigma 2p_z^2)(\pi 2p_x^2\ \pi 2p_y^2)(\pi * 2p_x^2\ \pi * 2p_y^2)$

Electronic configuration of O_2^-

$(\sigma 1s^2)(\sigma * 1s^2)(\sigma 2s^2)(\sigma * 2s^2)(\sigma 2p_z^2)(\pi 2p_x^2\ \pi 2p_y^2)(\pi * 2p_x^2\ \pi * 2p_y^1)$

superoxide is paramagnetic due to 1 unpaired electron.

20. (b) Gypsum $(CaSO_4 \cdot 2H_2O)$ is used to enhance setting time in portland cement.

21. (b) The balanced chemical reaction is :

$$2BeCl_2 + LiAlH_4 \longrightarrow 2BeH_2 + LiCl + AlCl_3$$

22. (d) The common method for the preparation of hydrogen peroxide is reaction with BaO_2 and H_2SO_4.

$$BaO_2 + H_2SO_4 \longrightarrow BaSO_4 + H_2O_2$$

23. (a) For a given metal $\Delta_f H^\circ$ always becomes less negative from fluoride to iodide.

24. (d) On moving down the group in alkali metals melting point decreases.

25. (a) The better method of preparation of BeF_2 is heating $(NH_4)_2BeF_4$

$$(NH_4)_2BeF_4 \xrightarrow{\Delta} BeF_2 + NH_4F$$

26. (a) Sodium superoxide is not stable.

27. (b) In solid state $BeCl_2$ exists as polymer, in vapour state it form chloro-bridged dimer while above 1200 K it exist as monomer.

28. (c) s-block elements are highly reactive and found in combined state.

B. Objective Questions

[One or more than one option(s) is/are correct]

1. (a, b) $6Na + \underset{\text{(from air)}}{2O_2} \xrightarrow{\text{combustion}} 2Na_2O + Na_2O_2$

C. Statement and Explanation Type Problems

1. (a) $M + (x+y)NH_3 \longrightarrow [M(NH_3)_x]^+ + [e(NH_3)_y]^-$

❑❑❑

CHAPTER

5

Extraction of Elements & Metallurgy

A. Objective Questions

[Only one option is correct]

1. Oxidation states of the metal in the minerals haematite and magnetite, respectively, are : **(IIT 2011)**

(a) II, III in haematite and III in magnetite (b) II, III in haematite and II in magnetite

(c) II in haematite and II, III in magnetite (d) III in haematite and II, III in magnetite

2. In the cyanide extraction process of silver from argentite ore, the oxidizing and reducing agents used are : **(IIT 2012)**

(a) O_2 and CO respectively (b) O_2 and Zn dust respectively

(c) HNO_3 and Zn dust respectively (d) HNO_3 and CO respectively

3. Sulphide ores are common for the metals : **[JEE (Advanced) 2013]**

(a) Ag, Cu and Pb (b) Ag, Cu and Sn (c) Ag, Mg and Pb (d) Al, Cu and Pb

4. The metal that cannot be obtained by electrolysis of an aqueous solution of its salts is : **[JEE (Mains) 2014]**

(a) Ag (b) Ca (c) Cu (d) Cr

5. In the context of the Hall-Heroult process for the extraction of Al, which of the following statements is false ? **[JEE (Mains) 2015]**

(a) Al^{3+} is reduced at the cathode to form Al.

(b) Na_3AlF_6 serves as the electrolyte.

(c) CO and CO_2 are produced in this process.

(d) Al_2O_3 is mixed with CaF_2 which lowers the melting point of the mixture and brings conductivity.

6. Which one of the following ores is best concentrated by froth floatation method ? **[JEE (Mains) 2016]**

(a) Magnetite (b) Siderite (c) Galena (d) Malachite

7. With respect to ore, Ellingham diagram helps to predict the feasibility of its : **[JEE (Mains) 2019]**

(a) zone refining (b) thermal reduction (c) electrolysis (d) vapour phase refining

8. The Mond process is used for the : **[JEE (Mains) 2019]**
(a) purification of Ni (b) purification of Zr and Ti
(c) extraction of Zn (d) extraction of Mo

9. Calamine, malachite, magnetite and cryolite, respectively, are : **[JEE (Advanced) 2019]**
(a) $ZnSO_4$, $Cu(OH)_2$, Fe_3O_4, Na_3AlF_6
(b) $ZnCO_3$, $CuCO_3$, Fe_2O_3, Na_3AlF_6
(c) $ZnSO_4$, $CuCO_3$, Fe_2O_3, AlF_3
(d) $ZnCO_3$, $CuCO_3$, $Cu(OH)_2$, Fe_3O_4, Na_3AlF_6

10. The purest form of commercial iron is: **[JEE Main (Jan.) 2020]**
(a) scrap iron and pig iron (b) wrought iron
(c) cast iron (d) pig iron

11. Among the reactions (a)-(d), the reaction(s) that does/do not occur in the blast furnace during the extraction of iron is/are: **[JEE Main (Jan.) 2020]**
(A) $CaO + SiO_2 \longrightarrow CaSiO_3$ (B) $3Fe_2O_3 + CO \longrightarrow 2Fe_3O_4 + CO_2$
(C) $FeO + SiO_2 \longrightarrow FeSiO_3$ (D) $FeO \longrightarrow Fe + \frac{1}{2}O_2$
(a) (C) and (D) (b) (A)
(c) (A) and (D) (d) (D)

12. Cast iron is used for the manufacture of: **[JEE Main (Sept.) 2020]**
(a) pig iron, scrap iron and steel (b) wrought iron and steel
(c) wrought iron and pig iron (d) wrought iron, pig iron and steel

13. According to the following diagram, *A* reduces BO_2 when the temperature is: **[JEE Main (Jan.) 2020]**

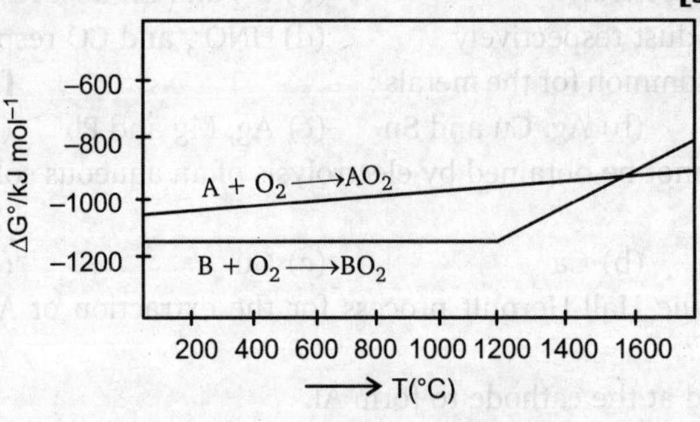

(a) > 1400 °C (b) < 1400 °C
(c) < 1200 °C (d) > 1200 °C but < 1400 °C

14. Among statements (A)–(D), the correct ones are: **[JEE Main (Sept.) 2020]**
(A) Lime stone is decomposed to CaO during the extraction of iron from its oxides.
(B) In the extraction of silver, silver is extracted as an anionic complex.
(C) Nickel is purified by Mond's process.
(D) Zr and Ti are purified by Van Arkel method.
(a) (A), (B), (C) and (D) (b) (C) and (D) only
(c) (A), (C) and (D) only (d) (B), (C) and (D) only

15. Match list-I with list-II :

List-I		List-II	
(A)	Mercury	(i)	Vapour phase refining
(B)	Copper	(ii)	Distillation refining
(C)	Silicon	(iii)	Electrolytic refining
(D)	Nickel	(iv)	Zone refining

Choose the most appropriate answer from the option given below :

[JEE Main (March) 2021]

(a) A-i, B-iv, C-ii, D-iii
(b) A-ii, B-iii, C-i, D-iv
(c) A-ii, B-iii, C-iv, D-i
(d) A-ii, B-iv, C-iii, D-i

16. Match List-I with List-II.

	List-I (Metal)		List-II (Ores)
(A)	Aluminium	(i)	Siderite
(B)	Iron	(ii)	Calamine
(C)	Copper	(iii)	Kaolinite
(D)	Zinc	(iv)	Malachite

Choose the correct answer from the options given below : **[JEE Main (Feb.) 2021]**

(a) (A) → (iv), (B) → (iii), (C) → (ii), (D) → (i)
(b) (A) → (ii), (B) → (iv), (C) → (i), (D) → (iii)
(c) (A) → (i), (B) → (ii), (C) → (iii), (D) → (iv)
(d) (A) → (iii), (B) → (i), (C) → (iv), (D) → (ii)

17. Match List-I with List-II.

	List-I		List-II
(A)	Siderite	(i)	Cu
(B)	Calamine	(ii)	Ca
(C)	Malachite	(iii)	Fe
(D)	Cryolite	(iv)	Al
		(v)	Zn

Choose the correct answer from the options given below : **[JEE Main (Feb.) 2021]**

(a) (A) → (iii), (B) → (i), (C) → (v), (D) → (ii)
(b) (A) → (i), (B) → (ii), (C) → (v), (D) → (iii)

(c) (A) → (iii), (B) → (v), (C) → (i), (D) → (iv)
(d) (A) → (i), (B) → (ii), (C) → (iii), (D) → (iv)

18. Match List-I and List-II : **[JEE Main (March) 2021]**

	List-I		List-II
(A)	Haematite	(i)	$Al_2O_3 \cdot xH_2O$
(B)	Bauxite	(ii)	Fe_2O_3
(C)	Magnetite	(iii)	$CuCO_3 \cdot Cu(OH)_2$
(D)	Malachite	(iv)	Fe_3O_4

Choose the correct answer from the options given below:

(a) (A)-(ii), (B)-(iii), (C)-(i), (D)-(iv)
(b) (A)-(iv), (B)-(i), (C)-(ii), (D)-(iii)
(c) (A)-(i), (B)-(iii), (C)-(ii), (D)-(iv)
(d) (A)-(ii), (B)-(i), (C)-(iv), (D)-(iii)

19. Ellingham diagram is a graphical representation of: **[JEE Main (Feb.) 2021]**

(a) ΔH *vs* T
(b) ΔG *vs* T
(c) ΔG *vs* P
(d) $(\Delta G - T\Delta S)$ *vs* T

20. The point of intersection and sudden increase in the slope, in the diagram given below, respectively, indicates: **[JEE Main (March) 2021]**

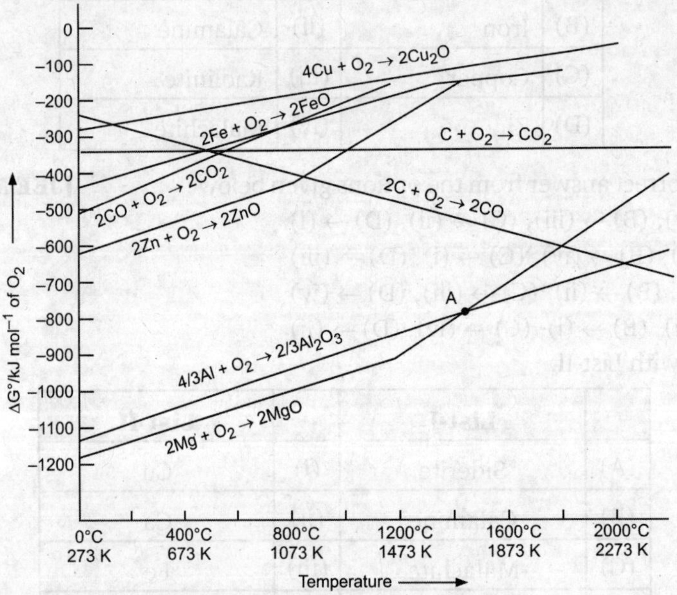

(a) $\Delta G = 0$ and melting or boiling point of the metal oxide
(b) $\Delta G > 0$ and decomposition of the metal oxide
(c) $\Delta G < 0$ and decomposition of the metal oxide
(d) $\Delta G = 0$ and reduction of the metal oxide

21. The method used for the purification of Indium is : **[JEE Main (Feb.) 2021]**

(a) Van Arkel method
(b) liquation
(c) zone refining
(d) vapour phase refining

22. In isolation of which one of the following metals from their ores, the use of cyanide salt is not commonly involved? **[JEE Main (June) 2022 (I)]**

(a) Zinc (b) Gold (c) Silver (d) Copper

23. In the metallurgical extraction of copper, following reaction is used :

$$FeO + SiO_2 \longrightarrow FeSiO_3$$ **[JEE Main (June) 2022 (II)]**

FeO and $FeSiO_3$ respectively are :

(a) gangue and flux (b) flux and slag
(c) slag and flux (d) gangue and slag

24. Refining using liquation method is the most suitable for metals with : **[JEE Main (July) 2022 (I)]**

(a) Low melting point
(b) High boiling point
(c) High electrical conductivity
(d) Less tendency to be soluble in melts than impurities

25. The major component of which of the following ore is sulphide based mineral? **[JEE Main (Jan.) 2023 (II)]**

(a) Calamine (b) Siderite (c) Sphalerite (d) Malachite

26. Which of the following is used as a stabilizer during the concentration of sulphide ores? **[JEE Main (April) 2023 (I)]**

(a) Pine oils (b) Xanthates (c) Fatty acids (d) Cresols

27. Which of the following metals can be extracted through alkali leaching technique? **[JEE Main (April) 2023 (I)]**

(a) Cu (b) Sn
(c) Pb (d) Au

28. The reaction representing the Mond process for metal refining is ________. **[JEE Main (Jan.) 2023 (I)]**

(a) $Ni + 4CO \xrightarrow{\Delta} Ni(CO)_4$

(b) $2K[Au(CN)_2] + Zn \xrightarrow{\Delta} K_2[Zn(CN)_4] + 2Au$

(c) $Zn + 2I_2 \xrightarrow{\Delta} ZrI_4$

(d) $ZnO + C \xrightarrow{\Delta} Zn + CO$

29. Identify the incorrect pair from the following : **[JEE Main (Jan.) 2024]**

(a) Fluorspar-BF_3 (b) Cryolite-Na_3AlF_6
(c) Fluoroapatite-$3Ca_3(PO_4)_2 \cdot CaF_2$ (d) Carnallite-$KCl \cdot MgCl_2 \cdot 6H_2O$

B. Objective Questions

[One or more than one option(s) is/are correct]

1. Extraction of metal from the ore **cassiterite** involves : **(IIT 2011)**
   (a) carbon reduction of an oxide ore (b) self-reduction of a sulphide ore
   (c) removal of copper impurity (d) removal of iron impurity

2. Upon heating with Cu_2S, the reagent(s) that give copper metal is/are :
   [JEE (Advanced) 2014]
   (a) $CuFeS_2$ (b) CuO (c) Cu_2O (d) $CuSO_4$

3. Copper is purified by electrolytic refining of blister copper. The correct statement(s) about this process is(are) : **[JEE (Advanced) 2015]**
   (a) Impure Cu strip is used as cathode
   (b) Acidified aqueous $CuSO_4$ is used as electrolyte
   (c) Pure Cu deposits at cathode
   (d) Impurities settle as anode-mud

4. Extraction of copper from copper pyrite ($CuFeS_2$) involves : **[JEE (Advanced) 2016]**
   (a) crushing followed by concentration of the ore by froth-floatation
   (b) removal of iron as slag
   (c) self-reduction step to produce 'blister copper' following evolution of SO_2
   (d) refining of 'blister copper' by carbon reduction

5. The cyanide process of gold extraction involves leaching out gold from its ore with CN^- in the presence of Q in water to form R. Subsequently, R is treated with T to obtain Au and Z. Choose the correct option(s). **[JEE (Advanced) 2019]**
   (a) R is $[Au(CN)_4]^-$. (b) T is Zn.
   (c) Q is O_2. (d) Z is $[Zn(CN)_4]^{2-}$.

6. Which among the following statement(s) is (are) true for the extraction of aluminium from bauxite? **[JEE (Advanced) 2020]**
   (a) Hydrated Al_2O_3 precipitates, when CO_2 is bubbled through a solution of sodium aluminate.
   (b) Addition of Na_3AlF_6 lowers the melting point of alumina.
   (c) CO_2 is evolved at the anode during electrolysis.
   (d) The cathode is a steel vessel with a lining of carbon.

C. Matching Type Problems

1. Match the extraction processes listed in column I with metals listed in column II.
   (IIT 2006)

	Column I		Column II
(a)	Self reduction	(p)	Lead
(b)	Carbon reduction	(q)	Silver

(c)	Complex formation and displacement by metal	(r)	Copper
(d)	Decomposition of iodide	(s)	Boron

2. Match the conversions in column I with the type(s) of reaction(s) given in column II.

(IIT 2008)

	Column I		Column II
(a)	$PbS \longrightarrow PbO$	(p)	Roasting
(b)	$CaCO_3 \longrightarrow CaO$	(q)	Calcination
(c)	$ZnS \longrightarrow Zn$	(r)	Carbon reduction
(d)	$Cu_2S \longrightarrow Cu$	(s)	Self reduction

3. Match the anionic species given in Column-I that are present in the ore(s) given in Column-II:

[JEE (Advanced) 2015]

	Column-I		Column-II
(a)	Carbonate	(p)	Siderite
(b)	Sulphide	(q)	Malachite
(c)	Hydroxide	(r)	Bauxite
(d)	Oxide	(s)	Calamine
		(t)	Argentite

4. Match List I with List II : (Both having metallurgical terms) **[JEE Main (July) 2021 (II)]**

	List I		List II
(A)	Concentration of Ag ore	(i)	Reverberatory furnace
(B)	Blast furnace	(ii)	Pig iron
(C)	Blister copper	(iii)	Leaching with dilute NaCN solution
(D)	Froth floatation method	(iv)	Sulfide ores

Choose the correct answer from the options given below :

(a) (A)–(iii), (B)–(ii), (C)–(i), (D)–(iv)
(b) (A)–(iii), (B)–(iv), (C)–(i), (D)–(ii)
(c) (A)–(iv), (B)–(i), (C)–(iii), (D)–(ii)
(d) (A)–(iv), (B)–(iii), (C)–(ii), (D)–(i)

5. Match List I with List II : **[JEE Main (Jan.) 2023 (I)]**

	List I		List II
(A)	Reverberatory furnace	I.	Pig Iron
(B)	Electrolytic cell	II.	Aluminium
(C)	Blast furnace	III.	Silicon
(D)	Zone refining furnace	IV.	Copper

(a) A–IV, B–II, C–I, D–III
(b) A–I, B–IV, C–II, D–III
(c) A–I, B–III, C–II, D–IV
(d) A–III, B–IV, C–I, D–II

D. Passage Based Problems

Passage-1

Copper is the most noble of the first transition metals and occurs in small deposits in several countries. Ores of copper include chalcanthite ($CuSO_4 \cdot 5H_2O$), atacamite [$Cu_2Cl(OH)_3$], cuprite (Cu_2O), copper glance (Cu_2S) and malachite [$Cu_2(OH)_2CO_3$]. However, 80% of the world copper production comes from the ore of chalcopyrite ($CuFeS_2$). The extraction of copper from chalcopyrite involves partial roasting, removal of iron and self reduction.

(IIT 2010)

1. Partial roasting of chalcopyrite produces :
(a) Cu_2S and FeO (b) Cu_2O and FeO
(c) CuS and Fe_2O_3 (d) Cu_2O and Fe_2O_3

2. Iron is removed from chalcopyrite as :
(a) FeO (b) FeS (c) Fe_2O_3 (d) $FeSiO_3$

3. In self-reduction, the reducing species is :
(a) S (b) O^{2-} (c) S^{2-} (d) SO_2

Answers

[A] Objective Questions (Only one option is correct)

1. (d) **2.** (b) **3.** (a) **4.** (b) **5.** (b) **6.** (c) **7.** (b) **8.** (a) **9.** (d) **10.** (b)
11. (a) **12.** (b) **13.** (a) **14.** (a) **15.** (c) **16.** (d) **17.** (c) **18.** (d) **19.** (b) **20.** (a)
21. (c) **22.** (d) **23.** (d) **24.** (a) **25.** (d) **26.** (d) **27.** (d) **28.** (a) **29.** (a)

[B] Objective Questions (One or more than one option(s) is/are correct)

1. (a, c, d) **2.** (b, c, d) **3.** (b, c, d) **4.** (a, b, c) **5.** (b, c, d)
6. (a, b, c, d)

[C] Matching Type Problems

1. $a \to p, r;\ b \to p;\ c \to q;\ d \to s$
2. $a \to p;\ b \to q;\ c \to p, r;\ d \to p, r, s$
3. $a \to p, q, s;\ b \to t;\ c \to q, r;\ d \to r$
4. (a) (A)–(iii), (B)–(ii), (C)–(i), (D)–(iv)
5. (a) A–IV, B–II, C–I, D–III

[D] Passage Based Problems

Passage-1

1. (b) **2.** (d) **3.** (c)

Hints and Solutions

A. Objective Questions

[Only one option is correct]

1. (d) Haematite : $Fe_2O_3 : 2x + 3 \times (-2) = 0$

$$x = 3$$

Magnetite : Fe_3O_4 [an equimolar mixture of FeO and Fe_2O_3]

$FeO : x - 2 = 0 \Rightarrow x = 2$

$Fe_2O_3 :\ x = 3$

2. (b) $4Ag + 8NaCN + 2H_2O + O_2(\text{air}) \longrightarrow \underset{\text{sodium argento cyanide (soluble)}}{4NaAg(CN)_2} + 4NaOH$

By adding highly electropositive metal (*e.g.*, Zn) the Ag may be recovered.

$$2NaAg(CN)_2 + Zn \longrightarrow \underset{\text{Sodium zinco-cyanide}}{Na_2Zn(CN)_4} + 2Ag\downarrow$$

3. (a) Ag, Cu and Pb

$$Ag \longrightarrow Ag_2S$$
$$Cu \longrightarrow CuFeS_2$$
$$Pb \longrightarrow PbS$$

4. (b) On electrolysis only in case of Ca^{2+} salt aqueous solution H_2 gas discharge at Cathode.

Case of Cr

At cathode : $Cr^{3+} + 2e^- \longrightarrow Cr$

So, Cr is deposited.

Case of Ag

At cathode : $Ag^+ + e^- \longrightarrow Ag$

So, Ag is deposited.

Case of Cu

At cathode : $Cu^{2+} + 2e^- \longrightarrow Cu$

Case of Ca^{2+}

At cathode : $H_2O + e^- \longrightarrow \frac{1}{2}H_2 + OH^-$

5. (b) In Hall-Heroult process for the extraction of Al, electrolyte is Al_2O_3 dissolved in Na_3AlF_6 containing a little of CaF_2.

6. (c) Froth floatation method is suitable for sulphide ores thus, PbS *i.e.*, galena is best concentrated by this method.

7. (b) Ellingham diagrams help us in predicting the feasibility of thermal reduction of an ore.

8. (a)

9. (d)

10. (b) Purest form of commercial iron is wrought iron.

11. (a) Following reaction takes place in blast furnace during extraction of iron.

$$CaO + SiO_2 \longrightarrow CaSiO_3$$
$$3Fe_2O_3 + CO \longrightarrow 2Fe_3O_4 + CO_2$$

12. (b) Cast iron is used for the manufacture of wrought iron and steel.

13. (a) A reduces BO_2

$$A + BO_2 \longrightarrow AO_2 + B$$

if

$$A + O_2 \longrightarrow AO_2 = \Delta G_1^\circ$$
$$B + O_2 \longrightarrow BO_2 = \Delta G_2^\circ$$
$$\Delta_r G^\circ = \Delta G_1^\circ - \Delta G_2^\circ$$

For reaction to be spontaneous

$$\Delta_r G^\circ < 0 \quad \Delta G_1^\circ < \Delta G_2^\circ \qquad \text{...(i)}$$

Condition (i) will valid from graph above 1400°C temperature.

14. (a) In extraction of iron, lime stone ($CaCO_3$) is used as a flux. Which is decomposed to CaO in blast furnace.

$$CaCO_3(s) \longrightarrow CaO(s) + CO_2(g)$$

In extraction of silver

$$Ag + NaCN \longrightarrow \underset{\text{Anionic complex}}{[Ag(CN)_2]^-}$$

Nickel is purified by Mond's process.
Zr and Ti are purified by Van Arkel method.

15. (c)

Element	**Refining method**
Mercury	Distillation refining
Copper	Electrolytic refining
Silicon	Zone refining
Nickel	Vapour phase refining

16. (d) Siderite → $FeCO_3$ (iron ore)
Calamine → $ZnCO_3$ (zinc ore)
Kaolinite → $Al_2(OH)_4 \cdot Si_2O_5$ (Aluminium ore)
Malachite → $Cu(OH)_2 \cdot CuCO_3$ (Copper ore)

17. (c)

Ore	Formula	Metal present
Siderite	$FeCO_3$	Fe
Calamine	$ZnCO_3$	Zn
Malachite	$Cu(OH)_2 \cdot CuCO_3$	Cu
Cryolite	Na_3AlF_6	Al

18. (d)

Ore	Formula
Haematite	Fe_2O_3
Bauxite	$Al_2O_3 \cdot xH_2O$
Magnetite	Fe_3O_4
Malachite	$CuCO_3 \cdot Cu(OH)_2$

19. (b) Ellingham diagram is a graphical representation of ΔG *vs* T for various metal oxides.

20. (a) At intersection point $\Delta G = 0$ and sudden increase in slope is due to melting or boiling point of the metal oxide.

21. (c) Ga, In, Si and Ge are refined by 'zone refining' process.

22. (d) For Zn $\longrightarrow$ KCN is used. Ag/Au leached by cyanide process.

$$4Au + 8NaCN + O_2 + 2H_2O \longrightarrow 4Na[Au(CN)_2] + 4NaOH$$

23. (d) Gangue are impurities present in the metal ore.
Flux → chemical substance used to convert gangue into easily removable form.
Slag → gangue and flux combine to form slag.

24. (a) Liquation is used for the purification of metals having low melting point.

25. (d) Calamine : $ZnCO_3$; Siderite : $FeCO_3$;
Sphalerite : ZnS ; Malachite : $CuCO_3 \cdot Cu(OH)_2$

26. (d) Cresol is used as stabilizer.

27. (d) Leaching is employed for Ag, Au, Al and for some low grade Cu ores.

28. (a) Mond's process :

$$Ni + 4CO \xrightarrow{\Delta} [Ni(CO)_4]$$

29. (a) Fluorspar is CaF_2.

B. Objective Questions

[One or more than one option(s) is/are correct]

1. (a, c, d) $$SnO_2 + 2C \longrightarrow 2CO + Sn$$

The ore cassiterite contains the impurity of Fe, Mn, W and traces of Cu.

2. (b, c, d)

(a) $$2CuFeS_2 + O_2 \xrightarrow{\Delta} Cu_2S + 2FeS + SO_2$$

(b) $$4CuO \xrightarrow{1100°C} 2Cu_2O + O_2$$

$$2Cu_2O + Cu_2S \xrightarrow{\Delta} 6Cu + O_2$$

(c) $$Cu_2S + 2Cu_2O \xrightarrow{\Delta} 6Cu + SO_2$$

(d) $$CuSO_4 \xrightarrow{720°C} CuO + SO_2 + \frac{1}{2}O_2$$

$$4CuO \xrightarrow{1100°C} 2Cu_2O + O_2$$

$$2Cu_2O + Cu_2S \xrightarrow{\Delta} 6Cu + SO_2$$

3. (b, c, d)

(a) Impure Cu strip is used as anode and impurities settle as anode mud.
(b) Pure Cu deposits at cathode.
(c) Acidified aqueous $CuSO_4$ is used as electrolyte.

At anode : $$Cu_{(s)} \longrightarrow Cu^{2}_{(aq.)} + 2e^-$$

At cathode : $$Cu^{2+}_{(aq.)} + 2e^- \longrightarrow Cu_{(s)}$$

4. (a, b, c) $CuFeS_2$

(a) The ore contains sulphide ions so in the purification, concentration is done by froth-floatation process.

(b) $$2CuFeS_2 + O_2 \longrightarrow Cu_2S + 2FeS + SO_2$$

$$2FeS + 3O_2 \longrightarrow 2FeO + 2SO_2$$

$$FeO + SiO_2 \longrightarrow \underset{\text{(Slag)}}{FeSiO_3}$$

(c) $2Cu_2S + 3O_2 \longrightarrow 2Cu_2O + 2SO_2$

(Partial oxidation)

$$2Cu_2O + Cu_2S \longrightarrow \underset{\text{(Blister copper)}}{6Cu} + SO_2$$

(Self Reduction)

(d) Copper is refined by electrolytic refining.

5. (b, c, d)

$$Au \xrightarrow{NaCN+O_2} \underset{(R)}{[Au(CN)_2]^{\ominus}} \xrightarrow{Zn(T)} \underset{(Z)}{[Zn(CN)_4]^{2-}} + Au$$

6. (a, b, c, d)

(a) $2Na[Al(OH)_4]_{(aq.)} + CO_2 \longrightarrow Na_2CO_3 + H_2O + 2Al(OH)_3(\downarrow)$ or $Al_2O_3.2H_2O$ (ppt.)

(b) Function of Na_3AlF_6 is to lower the melting point of electrolyte.

(c) During electrolysis of Al_2O_3, the reactions at anode are :

$$[2Al^{3+}(l) + 3O^{2-}(l) \xrightarrow{\text{At anode}} O_2(gas) + 2e^-]$$

$$C(graphite) + O_2 \longrightarrow CO(\uparrow) + CO_2(\uparrow)$$

(d) The steel vessel with a lining of carbon acts as cathode.

C. Matching Type Problems

1. a — p, r; b — p; c — q; d — s

(a) $2PbO + PbS \longrightarrow 3Pb + SO_2$
$2Cu_2O + Cu_2S \longrightarrow 6Cu + SO_2$ } (Self reduction)

(b) $PbO + C \longrightarrow Pb + CO$

(c) $2Ag + 4CN + H_2O + \frac{1}{2}O_2 \longrightarrow 2[Ag(CN)_2] + 2OH$

$$2[Ag(CN)_2]^- + Zn \longrightarrow [Zn(CN)_4]^{2-} + 2Ag$$

(d) $2BI_3 \longrightarrow 2B + 3I_2$

2. a — p; b — q; c — p, r; d — p, r, s

(a) $PbS \xrightarrow{O_2/\Delta} PbO$ (Roasting of sulphide ore)

(b) $CaCO_3 \xrightarrow{\Delta} CaO + CO_2$ (Calcination of carbonate ore)

(c) $ZnS \xrightarrow[\text{Roasting}]{} ZnO \xrightarrow[-CO]{C} Zn$ (Reduction by carbon)

(d) $Cu_2S + 2Cu_2O \longrightarrow 6Cu + SO_2$ (Self reduction)

3. a → p, q, s; b → t; c → q, r; d → r

Siderite	$FeCO_3$
Malachite	$CuCO_3 \cdot Cu(OH)_2$
Bauxite	$AlO_x(OH)_{3-2x}$; $0 < x < 1$
Calamine	$ZnCO_3$
Argentite	Ag_2S

4. (a) (A)–(iii), (B)–(ii), (C)–(i), (D)–(iv)

5. (a) A–IV, B–II, C–I, D–III

D. Passage Based Problems

Passage-1

1. (b) Partial roasting of chalcopyrites involves the following reactions :

$$2CuFeS_2 + O_2 \longrightarrow Cu_2S + 2FeS + SO_2 \uparrow$$
$$2Cu_2S + 3O_2 \longrightarrow 2Cu_2O + 2SO_2 \uparrow$$
$$2FeS + 3O_2 \longrightarrow 2FeO + 2SO_2 \uparrow$$

2. (d)
$$FeO + SiO_2 \longrightarrow \underset{\text{(Slag)}}{FeSiO_3}$$

3. (c)
$$Cu_2S + 2Cu_2O \longrightarrow \underset{\text{(Blister copper)}}{6\,Cu} + SO_2 \uparrow$$

In the oxidation of $S^{2-} \longrightarrow S^{4+}$

S^{2-} is reducing agent.

□□□

CHAPTER

6

p-block Elements

A. Objective Questions

[Only one option is correct]

Group : 13 & 14

1. The increasing order of atomic radii of the following group 13 elements is: **[JEE (Advanced) 2016]**
   (a) Al<Ga<In<Tl (b) Ga<Al<In<Tl (c) Al<In<Ga<Tl (d) Al<Ga<Tl<In
2. The electronegativity of aluminium is similar to : **[JEE Main 2019]**
   (a) boron (b) lithium (c) beryllium (d) carbon
3. The relative stability of +1 oxidation state of group 13 elements follows the order: **[JEE Main 2019]**
   (a) Ga < Al < In < Tl (b) Tl < In < Ga < Al
   (c) Al < Ga < Tl < In (d) Al < Ga < In < Tl
4. Aluminium is usually found in +3 oxidation state. In contrast, thallium exists in +1 and +3 oxidation states. This is due to : **[JEE Main 2019]**
   (a) lanthanoid contraction (b) lattice effect
   (c) diagonal relationship (d) inert pair effect
5. The number of 2-centre-2-electron and 3-centre-2- electron bonds in B_2H_6 respectively, are: **[JEE Main 2019]**
   (a) 4 and 2 (b) 2 and 2 (c) 2 and 4 (d) 2 and 1
6. Diborane (B_2H_6) reacts independently with O_2 and H_2O to produce, respectively: **[JEE Main 2019]**
   (a) H_3BO_3 and B_2O_3 (b) B_2O_3 and H_3BO_3
   (c) B_2O_3 and $[BH_4]^-$ (d) HBO_2 and H_3BO_3
7. The element that shows greater ability to form $p\pi$-$p\pi$ multiple bonds, is: **[JEE Main 2019]**
   (a) C (b) Si (c) Ge (d) Sn
8. C_{60}, an allotrope of carbon contains: **[JEE Main 2019]**
   (a) 16 hexagons and 16 pentagons (b) 12 hexagons and 20 pentagons
   (c) 18 hexagons and 14 pentagons (d) 20 hexagons and 12 pentagons

9. The basic structural unit of feldspar, zeolites, mica, and asbestos is: **[JEE Main 2019]**

(a) $(SiO_3)^{2-}$ (b) SiO_2 (c) $(SiO_4)^{4-}$ (d) $\left(\begin{matrix} R \\ | \\ Si \\ | \\ R \end{matrix} - O \right)_n$

$(R = Me)$

10. In comparison to the zeolite process for the removal of permanent hardness, the synthetic resins method is: **[JEE Main (Jan.) 2020]**

(a) less efficient as it exchanges only anions
(b) more efficient as it can exchange only cations
(c) less efficient as the resin cannot be regenerated
(d) more efficient as it can exchange both cation as well as anions

11. The equation that represents the water – gas shift reactions is: **[JEE Main (Sept.) 2020]**

(a) $CO(g) + H_2O(g) \xrightarrow[\text{Catalyst}]{673K} CO_2(g) + H_2(g)$

(b) $C(s) + H_2O(g) \xrightarrow{1270K} CO(g) + H_2(g)$

(c) $2C(s) + O_2(g) + 4N_2(g) \xrightarrow{1273K} 2CO(g) + 4N_2(g)$

(d) $CH_4(g) + H_2O(g) \xrightarrow[\text{Ni}]{1273K} CO(g) + 3H_2(g)$

12. The reaction of $H_3N_3B_3Cl_3(A)$ with $LiBH_4$ in tetrahydrofuran gives inorganic benzene (B). Further, the reaction of (A) with (C) leads to $H_3N_3B_3(Me)_3$. Compounds (B) and (C) respectively, are: **[JEE Main (Jan.) 2020]**

(a) Borazine and MeBr (b) Boron nitride and MeBr
(c) Diborane and MeMgBr (d) Borazine and MeMgBr

13. Al_2O_3 was leached with alkali to get X. The solution of X on passing of gas Y, forms Z. X, Y and Z respectively are : **[JEE Main (Feb.) 2021]**

(a) $X = Na[Al(OH)_4], Y = SO_2, Z = Al_2O_3$
(b) $X = Na[Al(OH)_4], Y = CO_2, Z = Al_2O_3 \cdot xH_2O$
(c) $X = Al(OH)_3, Y = CO_2, Z = Al_2O_3$
(d) $X = Al(OH)_3, Y = SO_2, Z = Al_2O_3 . xH_2O$

14. Water does not produce CO on reacting with: **[JEE Main (Feb.) 2021]**

(a) CO_2 (b) C (c) CH_4 (d) C_3H_8

Group : 15, 16, 17 & 18

15. The reaction of P_4 with **X** leads selectively to P_4O_6. The **X** is : **(IIT 2009)**

(a) Dry O_2 (b) A mixture of O_2 and N_2
(c) Moist O_2 (d) O_2 in the presence of aqueous NaOH

16. Extra pure N_2 can be obtained by heating: **(IIT 2011)**

(a) NH_3 with CuO (b) NH_4NO_3
(c) $(NH_4)_2Cr_2O_7$ (d) $Ba(N_3)_2$

17. The shape of XeO_2F_2 molecule is : **(IIT 2012)**
(a) trigonal bipyramidal (b) square planar
(c) tetrahedral (d) see-saw

18. Which of the following is the wrong statement? **[JEE (Mains) 2013]**
(a) ONCl and ONO^- are not isoelectronic. (b) O_3 molecule is bent.
(c) Ozone is violet-black in solid state. (d) Ozone is diamagnetic gas.

19. Concentrated nitric acid, upon long standing, turns yellow-brown due to the formation of : **[JEE (Advanced) 2013]**
(a) NO (b) NO_2 (c) N_2O (d) N_2O_4

20. The correct statement for the molecule, CsI_3, is : **[JEE (Mains) 2014]**
(a) It is a covalent molecule (b) It contains Cs^+ and I_3^- ions
(c) It contains Cs^{3+} and I^- ions (d) It contains Cs^+, I^- and lattice I_2 molecule

21. Under ambient conditions, the total number of gases released as products in the final step of the reaction scheme shown below is: **[JEE (Advanced) 2014]**

$$XeF_6 \xrightarrow{\text{Complete Hydrolysis}} P + \text{other product}$$

$$P \xrightarrow{OH^-/H_2O} Q \xrightarrow{\text{Slow disproportionation in } OH^-/H_2O} \text{Products}$$

(a) 0 (b) 1 (c) 2 (d) 3

22. The product formed in the reaction of $SOCl_2$ with white phosphorous is: **[JEE (Advanced) 2014]**
(a) PCl_3 (b) SO_2Cl_2 (c) SCl_2 (d) $POCl_3$

23. Which among the following is the most reactive ? **[JEE (Mains) 2015]**
(a) I_2 (b) ICl (c) Cl_2 (d) Br_2

24. Which one has the highest boiling point ? **[JEE (Mains) 2015]**
(a) Kr (b) Xe (c) He (d) Ne

25. Under hydrolytic conditions, the compounds used for preparation of linear polymer and for chain termination, respectively, are : **[JEE (Advanced) 2015]**
(a) CH_3SiCl_3 and $Si(CH_3)_4$ (b) $(CH_3)_2SiCl_2$ and $(CH_3)_3SiCl$
(c) $(CH_3)_2SiCl_2$ and CH_3SiCl_3 (d) $SiCl_4$ and $(CH_3)_3SiCl$

26. The reaction of zinc with dilute and concentrated nitric acid, respectively produces : **[JEE (Mains) 2016]**
(a) N_2O and NO_2 (b) NO_2 and NO (c) NO and N_2O (d) NO_2 and N_2O

27. The pair in which phosphorus atoms have a formal oxidation state of +3 is : **[JEE (Mains) 2016]**
(a) orthophosphorous and pyrophosphorous acids
(b) pyrophosphorous and hypophosphoric acids
(c) orthophosphorous and hypophosphoric acids
(d) pyrophosphorous and pyrophosphoric acids

28. Which of the following reactions is an example of a redox reaction? **[JEE (Mains) 2017]**
(a) $XeF_6 + H_2O \longrightarrow XeOF_4 + 2HF$ (b) $XeF_6 + 2H_2O \longrightarrow XeO_2F_2 + 4HF$
(c) $XeF_4 + O_2F_2 \longrightarrow XeF_6 + O_2$ (d) $XeF_2 + PF_5 \longrightarrow [XeF]^+PF_6^-$

29. The products obtained when chlorine gas reacts with cold and dilute aqueous NaOH are: **[JEE (Mains) 2017]**

(a) Cl^- and ClO^- (b) Cl^- and ClO_2^-
(c) ClO^- and ClO_3^- (d) ClO_2^- and ClO_3^-

30. The compound that does not produce nitrogen gas by the thermal decomposition is: **[JEE (Mains) 2018]**

(a) NH_4NO_2 (b) $(NH_4)_2SO_4$
(c) $Ba(N_3)_2$ (d) $(NH_4)_2Cr_2O_7$

31. Which of the following are Lewis acids? **[JEE (Mains) 2018]**

(a) PH_3 and $SiCl_4$ (b) BCl_3 and $AlCl_3$
(c) PH_3 and BCl_3 (d) $AlCl_3$ and $SiCl_4$

32. Diborane (B_2H_6) reacts independently with O_2 and H_2O to produce, respectively : **[JEE (Mains) 2019]**

(a) HBO_2 and H_3BO_3 (b) B_2O_3 and H_3BO_3
(c) B_2O_3 and $[BH_4]^-$ (d) H_3BO_3 and B_2O_3

33. Which of the following liberates O_2 upon hydrolysis? **[JEE (Advanced) 2020]**

(a) Pb_3O_4 (b) KO_2 (c) Na_2O_2 (d) Li_2O_2

34. Which of the following polts is (are) correct for the given reaction? **[JEE (Advanced) 2020]**
($[P]_0$ is the initial concentration of P)

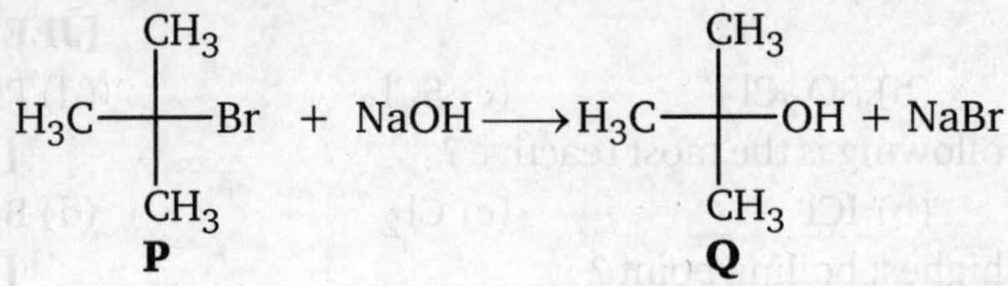

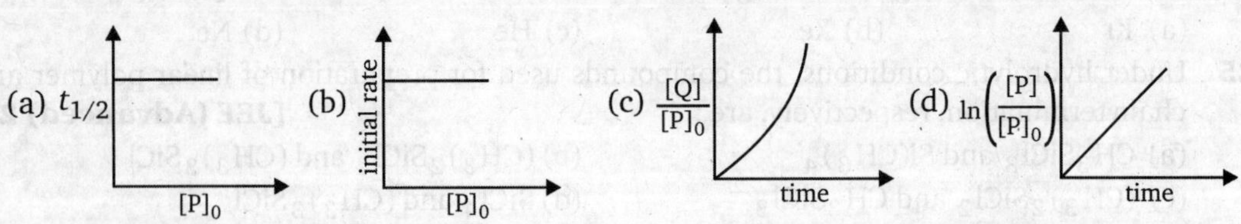

35. Aqua regia is used for dissolving noble metals (Au, Pt, etc.). The gas evolved in this process is: **[JEE Main (Sept.) 2020]**

(a) NO (b) N_2O_3 (c) N_2 (d) N_2O_5

36. On heating, lead (II) nitrate gives a brown gas (A). The gas (A) on cooling to a colourless solid/liquid (B). (B) on heating with NO changes to a blue solid (C). The oxidation number of nitrogen in solid (C) is : **[JEE Main (Sept.) 2020]**

(a) +4 (b) +3 (c) +2 (d) +5

37. White phosphorus on reaction with concentrated NaOH solution in an inert atmosphere of CO_2 gives phosphine and compound (X). (X) on acidification with HCl gives compound (Y). The basicity of compound (Y) is : **[JEE Main (Jan.) 2020]**

(a) 2 (b) 4 (c) 3 (d) 1

38. In a molecule of pyrophosphoric acid, the number of P—OH, P=O and P—O—P bonds/moiety(ies) respectively are: **[JEE Main (Sept.) 2020]**

(a) 4, 2 and 1
(b) 4, 2 and 0
(c) 2, 4 and 1
(d) 3, 3 and 3

39. Among statements (A)–(D), the correct ones are: **[JEE Main (Jan.) 2020]**

(A) decomposition of hydrogen peroxide gives dioxygen.
(B) like hydrogen peroxide, compounds such as $KClO_3$, $Pb(NO_3)_2$ and $NaNO_3$ when heated liberated dioxygen.
(C) 2-Ethylanthraquinone is useful for the industrial preparation of hydrogen peroxide
(D) Hydrogen peroxide is used for the manufacture of sodium perborate

(a) (A), (B), (C) and (D)
(b) (A) and (C) only
(c) (A), (B) and (C) only
(d) (A), (C) and (D) only

40. The number of bonds between sulphur and oxygen atoms in $S_2O_8^{2-}$ and the number of bonds between sulphur and sulphur atoms in rhombic sulphur, respectively, are: **[JEE Main (Jan.) 2020]**

(a) 4 and 8
(b) 4 and 6
(c) 8 and 6
(d) 8 and 8

41. In the following reactions, products (*A*) and (*B*) respectively are

$$\underset{\text{(Hot and conc.)}}{NaOH} + Cl_2 \longrightarrow (A) + \text{Side products}$$

$$\underset{\text{(Dry)}}{Ca(OH)_2} + Cl_2 \longrightarrow (B) + \text{Side products}$$

[JEE Main (Jan.) 2020]

(a) $NaClO_3$ and $Ca(OCl)_2$
(b) $NaOCl$ and $Ca(OCl)_2$
(c) $NaOCl$ and $Ca(ClO_3)_2$
(d) $NaClO_3$ and $Ca(ClO_3)_2$

42. The reaction in which hybridization of the underlined atom is affected is: **[JEE Main (Sept.) 2020]**

(a) $\underline{Xe}F_4 + SbF_5 \longrightarrow$
(b) $H_2\underline{S}O_4 + NaCl \xrightarrow{420K}$
(c) $H_3\underline{P}O_2 \xrightarrow{\text{Disproportionation}}$
(d) $\underline{N}H_3 \xrightarrow{H^+}$

43. Match List-I with List-II :

	List-I (Name of oxo acid)		List-II (Oxidation state of 'P')
(A)	Hypophosphorous acid	(i)	+5
(B)	Orthophosphoric acid	(ii)	+4
(C)	Hypophosphoric acid	(iii)	+3
(D)	Orthophosphorous acid	(iv)	+2
		(v)	+1

Choose the correct answer from the options given below : **[JEE Main (March) 2021]**

(a) (A)→(v), (B)→(i), (C)→(ii), (D)→(iii)
(b) (A)→(iv), (B)→(i), (C)→(ii), (D)→(iii)

(c) (A)→(iv), (B)→(v), (C)→(ii), (D)→(iii)
(d) (A)→(v), (B)→(iv), (C)→(ii), (D)→(iii)

44. (A) $HOCl + H_2O_2 \longrightarrow H_3O^+ + Cl^- + O_2$

(B) $I_2 + H_2O_2 + 2OH^- \longrightarrow 2I^- + 2H_2O + O_2^-$

Choose the correct option. **[JEE Main (Feb.) 2021]**

(a) H_2O_2 acts as reducing and oxidising agent respectively in equation (A) and (B)
(b) H_2O_2 acts as oxidising agent in equation (A) and (B)
(c) H_2O_2 acts as reducing agent in equation (A) and (B)
(d) H_2O_2 act as oxidizing and reducing agent respectively in equation (A) and (B)

45. Which oxoacid of phosphorus has the highest number of oxygen atoms present in its chemical formula? **[JEE Main (July) 2022 (I)]**

(a) Pyrophosphorous acid (b) Hypophosphoric acid
(c) Phosphoric acid (d) Pyrophosphoric acid

46. $A \xrightarrow{573\text{ K}} \text{Red phosphorus} \xrightarrow[\text{under pressure}]{\text{heat; 803 K}} B$

Red phosphorus is obtained by heating "*A*" at 573 K, and can be converted to "*B*" by heating at 803 K under pressure.

A and *B*, respectively, are : **[JEE Main (June) 2022 (I)]**

(a) β-black phosphorus and white phosphorus
(b) white phosphorus and β-black phosphorus
(c) α-black phosphorus and white phosphorus
(d) white phosphorus and α-black phosphorus

47. Which of the following oxoacids of sulphur contains "S" in two different oxidation states? **[JEE Main (June) 2022 (II)]**

(a) $H_2S_2O_3$ (b) $H_2S_2O_6$ (c) $H_2S_2O_7$ (d) $H_2S_2O_8$

48. Concentrated HNO_3 reacts with iodine to give : **[JEE Main (July) 2022 (II)]**

(a) HI, NO_2 and H_2O (b) HIO_2, N_2O and H_2O
(c) HIO_3, NO_2 and H_2O (d) HIO_4, N_2O and H_2O

49. The interhalogen compound formed from the reaction of bromine with excess of fluorine as a : **[JEE Main (July) 2022 (I)]**

(a) hypohalite (b) halate (c) perhalate (d) halite

50. Which of the phosphorus oxoacid can create silver mirror from $AgNO_3$ solution? **[JEE Main (Jan.) 2023 (I)]**

(a) $(HPO_3)_n$ (b) $H_4P_2O_5$ (c) $H_4P_2O_6$ (d) $H_4P_2O_7$

51. One mole of P_4 reacts with 8 moles of $SOCl_2$ to give 4 moles of *A*, *x* mole of SO_2 and 2 moles of *B*. *A*, *B* and *x* respectively are : **[JEE Main (April) 2023 (II)]**

(a) PCl_3, S_2Cl_2 and 4 (b) $POCl_3$, S_2Cl_2 and 4
(c) PCl_3, S_2Cl_2 and 2 (d) $POCl_3$, S_2Cl_2 and 2

52. Bond dissociation energy of *E*—H bond of the "H_2E" hydrides of group 16 elements (given below), follows order. **[JEE Main (Jan.) 2023 (II)]**

(A) O (B) S (C) Se (D) Te

(a) A > B > C > D (b) A > B > D > C (c) B > A > C > D (d) D > C > B > A

53. ClF_5 at room temperature is a : **[JEE Main (April) 2023 (I)]**
(a) Colourless gas with trigonal bipyramidal geometry
(b) Colourless gas with square pyramidal geometry
(c) Colourless liquid with square pyramidal geometry
(d) Colourless liquid was trigonal bipyramidal geometry

54. Which of the following compounds is an example of Freon : **[JEE Main (April) 2023 (II)]**
(a) $C_2Cl_2F_2$ (b) C_2HF_3 (c) $C_2H_2F_2$ (d) C_2F_4

55. For electron gain enthalpies of the elements denoted as $\Delta_{eg}H$, the incorrect option is : **[JEE Main (Feb.) 2023 (II)]**
(a) $\Delta_{eg}H(\text{Cl}) < \Delta_{eg}H(\text{F})$ (b) $\Delta_{eg}H(\text{Se}) < \Delta_{eg}H(\text{S})$
(c) $\Delta_{eg}H(\text{Te}) < \Delta_{eg}H(\text{Po})$ (d) $\Delta_{eg}H(\text{I}) < \Delta_{eg}H(\text{At})$

56. Inert gases have positive electron gain enthalpy. Its correct order is : **[JEE Main (Jan.) 2023 (I)]**
(a) Xe < Kr < Ne < He (b) He < Ne < Kr < Xe
(c) He < Xe < Kr < Ne (d) He < Kr < Xe < Ne

57. Choose the correct statements from the following :
A. All group 16 elements form oxides of general formula EO_2 and EO_3 when E = S, Se, Te and Po. Both the types of oxides are acidic in nature.
B. TeO_2 is an oxidising agent while SO_2 is reducing in nature.
C. The reducing property decreases from H_2S to H_2Te down the group.
D. The ozone molecule contains five lone pairs of electrons.
Choose the correct answer from the options given below : **[JEE Main (Jan.) 2024]**
(a) A and D only (b) B and C only (c) C and D only (d) A and B only

58. Given below are two statements :
Statement (I) : SiO_2 and GeO_2 are acidic while SnO and PbO are amphoteric in nature.
Statement (II) : Allotropic forms of carbon are due to property of catenation and $p\pi$-$d\pi$ bond formation.
In the light of the above statements, choose the most appropriate answer from the options given below : **[JEE Main (Jan.) 2024]**
(a) Both Statement I and Statement II are false
(b) Both Statement I and Statement II are true
(c) Statement I is true but Statement II is false
(d) Statement I is false but Statement II is true

59. The maximum covalency of a non-metallic group 15 element '*E*' with weakest *E* — *E* bond is : **[JEE Main (Jan.) 2025]**
(a) 5 (b) 3 (c) 6 (d) 4

60. The incorrect statements among the following is : **[JEE Main (Jan.) 2025]**
(a) PH_3 shows lower proton affinity than NH_3
(b) PF_3 exists but NF_5 does not

(c) NO_2 can dimerise easily

(d) SO_2 can act as an oxidizing agent but not as a reducing agent

61. Given below are the atomic numbers of some group 14 elements. The atomic number of the element with lowest melting point is : **[JEE Main (Jan.) 2025]**

(a) 14 (b) 6 (c) 82 (d) 50

62. The large difference between the melting and boiling points of oxygen and sulphur may be explained on the basis of : **[JEE Main (Jan.) 2025]**

(a) Atomic size (b) Atomicity

(c) Electronegativity (d) Electron gain enthalpy

63. Consider the following elements In, TI, Al, Pb, Sn and Ge.

The most stable oxidation states of elements with highest and lowest first ionisation enthalpies, respectively are : **[JEE Main (Jan.) 2025]**

(a) +2 and +3 (b) +4 and +3

(c) +4 and +1 (d) +1 and +4

B. Objective Questions

[One or more than one option(s) is/are correct]

1. When PbO_2 reacts with conc. HNO_3,the gas evolved is: **(IIT 2005)**

(a) NO_2 (b) O_2 (c) N_2 (d) N_2O

2. The reagent(s) used for softening the temporary hardness of water is(are) : **(IIT 2010)**

(a) $Ca_3(PO_4)_2$ (b) $Ca(OH)_2$

(c) $NaCO_3$ (d) NaOCl

3. Select the correct statement(s) for ortho boric acid : **[JEE (Advanced) 2014]**

(a) It behaves as a weak acid in water due to self ionization.

(b) Acidity of its aqueous solution increases upon addition for ethylene glycol.

(c) It has a three dimensional structure due to hydrogen bonding.

(d) It is a weak electrolyte in water.

4. The crystalline form of borax has: **[JEE (Advanced) 2016]**

(a) tetranuclear $[B_4O_5(OH)_4]^{2-}$ unit

(b) all boron atoms in the same plane

(c) equal number of sp^2 and sp^3 hybridized boron atoms

(d) one terminal hydroxide per boron atom.

5. Among the following, the correct statement (s) is (are): **[JEE (Advanced) 2017]**

(a) $Al(CH_3)_3$ has the three-centre two-electron bonds in its dimeric structure

(b) BH_3 has the three-centre two-electron bonds in its dimeric structure

(c) the Lewis acidity of BCl_3 is greater than that of $AlCl_3$

(d) $AlCl_3$ has the three-centre two-electron bonds in its dimeric structure.

6. Choose the correct statement(s) among the following. **[JEE (Advanced) 2020]**

(a) $SnCl_2.2H_2O$ is a reducing agent.

(b) SnO_2 reacts with KOH to form $K_2[Sn(OH)_6]$.

(c) A solution of $PbCl_2$ in HCl contains Pb^{2+} and Cl^- ions.

(d) The reaction of Pb_3O_4 with hot dilute nitric acid to give PbO_2 is a redox reaction.

7. The nitrogen oxide(s) that contain(s) N—N bond(s) is (are): **(IIT 2009)**

(a) N_2O (b) N_2O_3 (c) N_2O_4 (d) N_2O_5

8. With respect to graphite and diamond, which of the statement(s) given is(are) correct ? **(IIT 2012)**

(a) Graphite is harder than diamond.

(b) Graphite has higher electrical conductivity than diamond.

(c) Graphite has higher thermal conductivity than diamond.

(d) Graphite has higher C — C bond order than diamond.

9. Which of the following hydrogen halides react(s) with $AgNO_{3(aq.)}$ to give a precipitate that dissolves in $Na_2S_2O_{3(aq)}$? **(IIT 2012)**

(a) HCl (b) HF (c) HBr (d) HI

10. The pair(s) of reagents that yield paramagnetic species is/are : **[JEE (Advanced) 2014]**

(a) Na and excess of NH_3 (b) K and excess of O_2

(c) Cu and dilute HNO_3 (d) O_2 and 2-ethylanthraquinol

11. The correct statement(s) regarding, (i) HClO, (ii) $HClO_2$, (iii) $HClO_3$ and (iv) $HClO_4$, is (are) : **[JEE (Advanced) 2015]**

(a) The number of Cl $=$ O bonds in (ii) and (iii) together is two

(b) The number of lone pairs of electrons on Cl in (ii) and (iii) together is three

(c) The hybridization of Cl in (iv) is sp^3

(d) Amongst (i) to (iv), the strongest acid is (i)

12. The compound (s)with two lone pairs of electrons on the central atom is (are): **[JEE (Advanced) 2016]**

(a) BrF_5 (b) ClF_3 (c) XeF_4 (d) SF_4

13. The nitrogen containing compound produced in the reaction of HNO_3 with P_4O_{10}: **[JEE (Advanced) 2016]**

(a) can also be prepared by reaction of P_4 and HNO_3

(b) is diamagnetic

(c) contains one N—N bond

(d) reacts with Na metal producing brown gas

14. The colour of the X_2 molecules of group 17 elements changes gradually from yellow to violet down the group. This is due to: **[JEE (Advanced) 2017]**

(a) the physical state of X_2 at room temperature changes from gas to solid down the group.

(b) decrease in HOMO-LUMO gap down the group.

(c) decrease in $\pi^* - \sigma^*$ gap down the group.

(d) decrease in ionization energy down the group.

15. The correct statement(*s*) about the oxoacids, $HClO_4$ and HClO, is (are): **[JEE (Advanced) 2017]**

(a) the conjugate base of $HClO_4$ is weaker base than H_2O

(b) the central atom in both $HClO_4$ and HClO is sp^3 hybridized

(c) $HClO_4$ is formed in the reaction between Cl_2 and H_2O

(d) $HClO_4$ is more acidic than HClO because of the resonance stabilization of its anion.

16. The compound(s) which generate(s) N_2 gas upon thermal decomposition below 300°C is (are): **[JEE (Advanced) 2018]**

(a) NH_4NO_3 (b) $(NH_4)_2Cr_2O_7$

(c) $Ba(N_3)_2$ (d) Mg_3N_2

17. Based on the compounds of group 15 elements, the correct statement(s) is (are) : **[JEE (Advanced) 2018]**

(a) Bi_2O_5 is more basic than N_2O_5.

(b) NF_3 is more covalent than BiF_3.

(c) PH_3 boils at lower temperature than NH_3.

(d) The N—N single bond is stronger than the P—P single bond.

18. A tin chloride 'Q' undergoes the following reactions (not balanced)

$$Q + Cl^- \longrightarrow X$$
$$Q + Me_3N \longrightarrow Y$$
$$Q + CuCl_2 \longrightarrow Z + CuCl$$

X is monoanion having pyramidal geometry. Both Y and Z are neutral compounds. Choose the correct option(s): **[JEE (Advanced) 2019]**

(a) The central atom in X is sp^3 hybridized.

(b) There is a coordinate bond in Y.

(c) The oxidation state of the central atom in Z is +2.

(d) The central atom in Z has one lone pair of electrons.

19. With reference to aqua regia, choose the correct option(s). **[JEE (Advanced) 2019]**

(a) Reaction of gold with aqua regia produces NO_2 in the absence of air.

(b) The yellow colour of aqua regia is due to the presence of NOCl and Cl_2.

(c) Aqua regia is prepared by mixing conc. HCl and conc. HNO_3 in 3 : 1 (V/V) ratio.

(d) Reaction of gold with aqua regia produces an anion having Au in +3 oxidation state.

20. With respect to hypochlorite, chlorate and perchlorate ions, choose the correct statement(s). **[JEE (Advanced) 2020]**

(a) The hypochlorite ion is the strongest conjugate base.

(b) The molecular shape of only chlorate ion is influenced by the lone pair of electrons of Cl.

(c) The hypochlorite and chlorate ions disproportionate to give rise to identical set of ions.

(d) The hypochlorite ion oxidizes the sulphite ion.

C. Matching Type Problems

1. Match the reactions in **Column I** with the nature of the reaction/type of the products listed in **Column II.** **(IIT 2006)**

	Column I		Column II
(a)	$Bi^{3+} \longrightarrow [BiO]^+$	(p)	Heat
(b)	$[AlO_2]^- \longrightarrow Al(OH)_3$	(q)	Hydrolysis
(c)	$SiO_4^{4-} \longrightarrow Si_2O_7^{6-}$	(r)	Acidification
(d)	$(B_4O_7^{2-}) \longrightarrow [B(OH)_3]$	(s)	Dilution by water

2. Match each of the reactions given in **Column I** with the corresponding product(s) given in **Column II.** **(IIT 2009)**

	Column I		Column II
(a)	Cu + dil. HNO_3	(p)	NO
(b)	Cu + conc. HNO_3	(q)	NO_2
(c)	Zn + dil. HNO_3	(r)	N_2O
(d)	Zn + conc. HNO_3	(s)	$Cu(NO_3)_2$
		(t)	$Zn(NO_3)_2$

D. Passage Based Problems

Passage-1

The noble gases have closed-shell electronic configuration and are monoatomic gases under normal conditions. The low boiling points of the lighter noble gases are due to weak dispersion forces between the atoms and the absence of other interatomic interactions.

The direct reaction of xenon with fluorine leads to a series of compounds with oxidation numbers +2, +4 and +6. XeF_4 reacts violently with water to give XeO_3. The compounds of xenon exhibit rich stereochemistry and their geometries can be deduced considering the total number of electron pairs in the valence shell. **(IIT 2007)**

1. Argon is used in arc welding because of its :

(a) low reactivity with metal
(b) ability to lower the melting point of metal
(c) flammability
(d) high calorific value

2. The structure of XeO_3 is :

(a) linear
(b) plannar
(c) pyramidal
(d) T-shaped

3. XeF_4 and XeF_6 are expected to be :

(a) oxidising (b) reducing

(c) unreactive (d) strongly basic

Passage-2

There are some deposits of nitrates and phosphates in earth's crust. Nitrates are more soluble in water. Nitrates are difficult to reduce under the laboratory conditions but microbes do it easily. Ammonia forms large number of complexes with transition metal ions. Hybridisation easily explains the ease of sigma donation capability of NH_3 and PH_3. Phosphine is a flammable gas and is prepared from white phosphorus. **(IIT 2008)**

1. Among the following, the correct statement is :

(a) Phosphates have no biological significance in humans

(b) Between nitrates and phosphates, phosphates are less abundant in earth's crust

(c) Between nitrates and phosphates, nitrates are less abundant in earth's crust

(d) Oxidation of nitrates is possible in soil

2. Among the following, the correct statement is :

(a) Between NH_3 and PH_3, NH_3 is a better electron donor because the lone pair of electrons occupies spherical 's' orbital and is less directional.

(b) Between NH_3 and PH_3, PH_3 is a better electron donor because the lone pair of electrons occupies sp^3 orbital and is more directional

(c) Between NH_3 and PH_3, NH_3 is a better electron donor because the lone pair of electrons occupies sp^3 orbital and is more directional

(d) Between NH_3 and PH_3, PH_3 is a better electron donor because the lone pair of electrons occupies spherical 's' orbital and is less directional

3. White phosphorus on reaction with NaOH gives PH_3 as one of the products. This is a:

(a) dimerisation reaction (b) disproportionation reaction

(c) condensation reaction (d) precipitation reaction

Passage-3

Upon heating $KClO_3$ in the presence of catalytic amount of MnO_2, a gas *W* is formed. Excess amount of *W* reacts with white phosphorus to give *X*. The reaction of *X* with pure HNO_3 gives *Y* and *Z*. **[JEE (Advanced) 2017]**

1. *Y* and *Z* are, respectively:

(a) N_2O_5 and HPO_3 (b) N_2O_3 and H_3PO_4

(c) N_2O_4 and H_3PO_3 (d) N_2O_4 and HPO_3

2. *W* and *X* are, respectively:

(a) O_2 and P_4O_6 (b) O_2 and P_4O_{10}

(c) O_3 and P_4O_6 (d) O_3 and P_4O_{10}

E. Statement and Explanation Type Problems

Read the following questions and answer as per the direction given below :

(a) Statement 1 is true; Statement 2 is true; Statement 2 is a correct explanation of Statement 1.

(b) Statement 1 is true; Statement 2 is true; Statement 2 is not the correct explanation of Statement 1.

(c) Statement 1 is true; Statement 2 is false.

(d) Statement 1 is false; Statement 2 is true.

1. Statement 1 : Between $SiCl_4$ and CCl_4, only $SiCl_4$ reacts with water.
Statement 2 : $SiCl_4$ is ionic and CCl_4 is covalent. **(IIT 2001)**

2. Statement 1 : Boron always forms covalent bond.
Statement 2 : The small size of B^{3+} favours formation of covalent bond. **(IIT 2007)**

3. Statement 1 : In water, orthoboric acid behaves as a weak monobasic acid.
Statement 2 : In water, orthoboric acid acts as a proton donor. **(IIT 2007)**

4. Statement 1 : Pb^{4+} compounds are stronger oxidising agents than Sn^{4+} compounds.
Statement 2 : The higher oxidation states for the group 14 elements are more stable for the heavier members of the group due to 'inert pair effect'. **(IIT 2008)**

5. Statement 1 : Nitrogen and oxygen are the main components in the atmosphere but these do not react to form oxides of nitrogen.
Statement 2 : The reaction between nitrogen and oxygen requires high temperature.
[JEE (Mains) 2015]

6. Statement 1 : The first ionization energy of Pb is greater than of Sn.
Statement 2 : The first ionization energy of Ge is greater than that of Si.
[JEE Main (Jan.) 2025]

F. Integer Answer Type Problems

1. The coordination number of Al in the crystalline state of $AlCl_3$ is : **(IIT 2009)**
2. The value of n in the molecular formula $Be_nAl_2Si_6O_{18}$ is : **(IIT 2010)**
3. Among the following, the number of compounds which can react with PCl_5 to give $POCl_3$ is $O_2, CO_2, SO_2, H_2O, H_2SO_4, P_4O_{10}$: **(IIT 2011)**
4. Three moles of B_2H_6 are completely reacted with methanol. The number of moles of boron containing product formed is : **[JEE (Advanced) 2015]**
5. At 143 K, the reaction of XeF_4 with O_2F_2 produces a xenon compound *Y*. The total number of lone pair(s) of electrons present on the whole molecule of *Y* is ___. **[JEE (Advanced) 2019]**
6. Chlorine reacts with hot and concentrated NaOH and produces compounds (*X*) and (*Y*). Compound (*X*) gives white precipitate with silver nitrate solution. The average bond order between Cl and O atoms in (*Y*) is ______. **[JEE Main (Jan.) 2020 (I)]**
7. The number of halogen/(s) forming halic (V) acid is _________. **[JEE Main (Aug.) 2021 (I)]**

8. The reaction of white phosphorus on boiling with alkali in inert atmosphere resulted in the formation of product '*A*'. The reaction of 1 mol of '*A*' with excess of $AgNO_3$ in aqueous medium gives ________ mol(s) of Ag.
(Round off to the nearest integer). **[JEE Main (March) 2021 (I)]**

9. The number of non-ionisable protons present in the product *B* obtained from the following reaction is ________.

$$C_2H_5OH + PCl_3 \longrightarrow C_2H_5Cl + A$$
$$A + PCl_3 \longrightarrow B$$

[JEE Main (July) 2022 (II)]

10. The number of interhalogens from the following having square pyramidal structure is :
$ClF_3, IF_7, BrF_5, BrF_3, I_2Cl_6, IF_5, ClF, ClF_5$ **[JEE Main (July) 2022 (I)]**

11. The ratio of sigma and π-bonds present in pyrophosphoric acid is ________.
[JEE Main (April) 2023 (II)]

12. Sum of π-bonds present in peroxodisulphuric acid and pyrosulphuric acid is :
[JEE Main (Jan.) 2023 (II)]

13. The sum of lone pairs present on the central atom of the interhalogen IF_5 and IF_7 is ________.
[JEE Main (April) 2023 (I)]

14. Sum of oxidation states of bromine in bormic acid and perbromic acid is ________.
[JEE Main (Feb.) 2023 (I)]

15. XeF_4 reacts with SbF_5 to form
$[XeF_m]^{n+}[SbF_y]^{z-}$

$m + n + y + z =$ ________. **[JEE Main (April) 2023 (I)]**

16. Among the following oxide of *p*-block elements, number of oxides having amphoteric nature is
$Cl_2O_7, CO, PbO_2, N_2O, NO, Al_2O_3, SiO_2, N_2O_5, SnO_2$ **[JEE Main (Jan.) 2024]**

17. From the given list, the number of compounds with +4 oxidation state of sulphur ________.
$SO_3, H_2SO_3, SOCl_2, SF_4, BaSO_4, H_2S_2O_7$ **[JEE Main (Jan.) 2024]**

18. Total number of species from the following which can undergo disproportionation reaction ________.

$H_2O_2, ClO_3^-, P_4, Cl_2, Ag, Cu^{1+}, F_2, NO_2, K^+$ **[JEE Main (Jan.) 2024]**

19. A group 15 element forms $d\pi - d\pi$ bond with transition metals. It also forms hydride, which is a strongest base among the hydrides of other group members that form $d\pi - d\pi$ bond. The atomic number of the elements is .... **[JEE Main (Jan.) 2025]**

Answers

[A] Objective Questions (Only one option is correct)

1. (b) **2.** (c) **3.** (d) **4.** (d) **5.** (a) **6.** (b) **7.** (a) **8.** (d) **9.** (c) **10.** (d)
11. (a) **12.** (d) **13.** (b) **14.** (a) **15.** (b) **16.** (d) **17.** (d) **18.** (a) **19.** (b) **20.** (b)
21. (c) **22.** (a) **23.** (b) **24.** (b) **25.** (b) **26.** (a) **27.** (a) **28.** (c) **29.** (a) **30.** (b)
31. (b) **32.** (b) **33.** (b) **34.** (a) **35.** (a) **36.** (b) **37.** (d) **38.** (a) **39.** (a) **40.** (d)
41. (a) **42.** (a) **43.** (a) **44.** (c) **45.** (d) **46.** (d) **47.** (a) **48.** (c) **49.** (b) **50.** (b)
51. (a) **52.** (a) **53.** (c) **54.** (a) **55.** (b) **56.** (c) **57.** (d) **58.** (c) **59.** (d) **60.** (d)
61. (d) **62.** (b) **63.** (c)

[B] Objective Questions (One or more than one option(s) is/are correct)

1. (b) **2.** (b, c, d) **3.** (b, d) **4.** (a, d) **5.** (a, b, c)
6. (a, b, c) **7.** (a, b, c) **8.** (b, d) **9.** (a, c, d) **10.** (a, b, c)
11. (b, c) **12.** (b, c) **13.** (b, d) **14.** (b, c) **15.** (a, b, d)
16. (b, c) **17.** (a, b, c) **18.** (a, b) **19.** (b, c, d) **20.** (a, b, d)

[C] Matching Type Problems

1. $a \rightarrow q, s$; $b \rightarrow q, s$; $c \rightarrow r$; $d \rightarrow q, r$ **2.** $a \rightarrow p, s$; $b \rightarrow q, s$; $c \rightarrow r, t$; $d \rightarrow q, t$

[D] Passage Based Problems

Passage-1

1. (a) **2.** (c) **3.** (a)

Passage-2

1. (c) **2.** (c) **3.** (b)

Passage-3

1. (a) **2.** (b)

[E] Statement and Explanation Type Problems

1. (c) **2.** (a) **3.** (a) **4.** (c) **5.** (b)
6. (c)

[F] Integer Answer Type Problems

1. (2) **2.** (3) **3.** (4) **4.** (6) **5.** (19)
6. (1.67) **7.** (3) **8.** (4) **9.** (2) **10.** (3)
11. (6) **12.** (6) **13.** (1) **14.** (12) **15.** (11)
16. (3) **17.** (3) **18.** (6) **19.** (15)

Hints and Solutions

A. Objective Questions

[Only one option is correct]

Group : 13 & 14

1. (b) Ga is smaller due to poor shielding of *d*-orbital.

2. (c) Diagonal relationship exist between 'Be' and 'Al' due to similarity in properties. So there is similarity in electronegativity as well so

$$\text{E.N. of Be} = \text{E.N. of Al}$$

3. (d) For Group 13 element +1 oxidation state (lower oxidation state) stability increases down the group due to inert pair effect.

$$\text{Al} < \text{Ga} < \text{In} < \text{Tl}$$

4. (d) Due to inert pair effect down the group lower oxidation state become more stable.

5. (a)

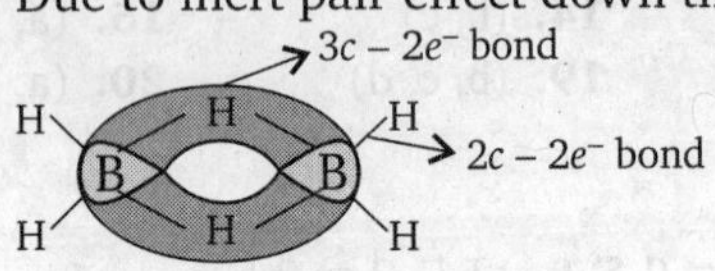

$2c - 2e^-$ bond = 4 bonds

$3c - 4e^-$ bond = 2 bonds

6. (b) $\text{Diborane}(B_2H_6) \xrightarrow{3O_2} B_2O_3 + 3H_2O$

(combustion of diborane)

$\text{Diborane}(B_2H_6) \xrightarrow{6H_2O} 2H_3BO_3 + 3H_2\uparrow$

(hydrolysis of diborane)

7. (a) $p\pi$-$p\pi$ bond in carbon are formed by $2p\pi$-$2p\pi$ bonding orbitals. Which is stronger compared to $3p\pi$-$3p\pi$ bonds in other elements.

8. (d) C_{60} an allotrope of carbon contains: 20 hexagons and 12 pentagons.

9. (c) Feldspar : $KAl_3Si_3O_8 - NaAlSi_3O_8 - CaAl_2Si_2O_8$

Zeolite : $NaAlSi_2O_6 \cdot H_2O$

Mica : $KAl_3Si_3O_{10}(OH)_2$

Asbestos : $Mg_3Si_2O_5(OH)_4$

These are silicates having basic unit $(SiO_4)^{4-}$.

10. (d) Synthetic resin method are more superior method as they remove all type of cation and anion and the resultant water is distilled water.

$$\underset{\text{Carboxylic acid}}{R-COO^-H^+} + \underset{\text{Calcium chloride}}{CaCl_2} \longrightarrow \underset{\text{Calcium salt of carboxylic acid}}{(RCOO)_2Ca} + 2H^+ + \underset{\text{Chloride ions}}{2Cl^-}$$

$$R-NH_3OH + \underset{\text{Sulphate ions}}{SO_4^{2-}} \longrightarrow (R-NH_3)_2SO_4^{2-} + \underset{\text{Hydroxide ions}}{2OH^-}$$

$$\underset{\text{Hydrogen ion}}{H^+} + \underset{\text{Hydroxide ion}}{OH^-} \longrightarrow \underset{\text{Water}}{H_2O}$$

11. (a) Water gas shift reaction

$$CO(g) + H_2O(g) \xrightarrow[\text{Catalyst}]{673K} CO_2(g) + H_2(g)$$

12. (d) $\underset{(A)}{H_3N_3B_3Cl_3} \xrightarrow[\text{THF}]{\text{LiAlH}_4} \underset{(B)}{B_3N_3H_6}$

$\downarrow$ (C) MeMgBr

$H_3N_3B_3(Me)_3$

$(B) \rightarrow$ Borazine ($B_3N_3H_6$) is also called as Inorganic Benzene.

$(C) \rightarrow$ MeMgBr.

13. (b) Reactions are :

$$Al_2O_3 + 2NaOH + 3H_2O \longrightarrow \underset{(X)}{2Na[Al(OH)_4]}$$

$$\underset{(X)}{2Na[Al(OH)_4]} + \underset{(Y)}{CO_2} \longrightarrow \underset{(Z)}{Al_2O_3.xH_2O} + 2NaHCO_3$$

So, X is $Na[Al(OH)_4]$

Y is $CO_2(g)$

and Z is $Al_2O_3.xH_2O$

14. (a)

$$CO_2 + H_2O \longrightarrow H_2CO_3$$

$$C + H_2O \longrightarrow CO + H_2$$

$$CH_4 + H_2O \xrightarrow[1270K]{Ni} CO + 3H_2$$

$$C_3H_8 + H_2O \xrightarrow[1270K]{Ni} 3CO + H_2$$

Group : 15, 16, 17 & 18

15. (b) $P_4 + 3O_2 \xrightarrow{\text{in presence of } N_2} P_4O_6$

N_2 is used to retard the oxidation of P_4O_6 further.

16. (d) $Ba(N_3)_2 \xrightarrow{\Delta} Ba + 3N_2$

17. (d) Xe is in sp^3d hybrid state in XeO_2F_2 with 1 lone pair of electrons.

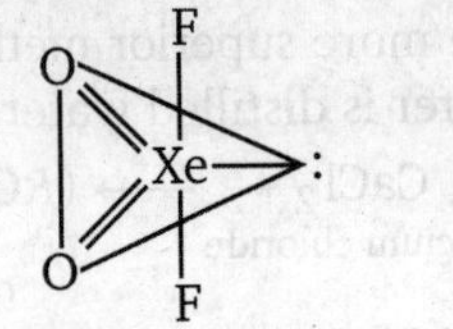

Geometry : Trigonal bipyramidal
Shape : See-saw

18. (a) ONCl and $ONCl^-$ are isoelectronic in nature.

19. (b) NO_2

$$4HNO_3(l) \xrightarrow{h\nu} 4NO_2(g)\uparrow + O_2(g)\uparrow + 2H_2O$$

20. (b) It contains Cs^+ and I_3^- ions.

21. (c) $XeF_6 + 3H_2O \longrightarrow XeO_3 + 3H_2F_2$

$$XeO_3 \longrightarrow HXeO_4^- \xrightarrow[\text{(Disproportionation)}]{OH^-/H_2O} XeO_{6(s)}^{4-} + Xe_{(g)} + H_2O_{(l)} + O_{2(g)}$$

22. (a) $P_{4(s)} + 8\,SOCl_{2(l)} \longrightarrow 4PCl_{3(l)} + 4\,SO_{2(g)} + 2\,S_2Cl_{2(g)}$

23. (b) Interhalogen compounds are generally more reactive than the halogens (except F_2).

24. (b) Boiling point increases down the group from He to Rn due to increase in van der Waals' forces of attraction as the size of the atom increases.

25. (b)

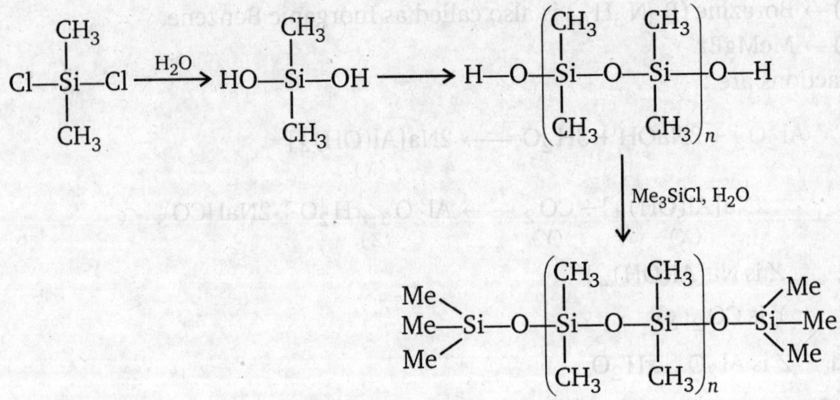

26. (a) $4Zn + 10HNO_{3(dil.)} \longrightarrow 4Zn(NO_3)_2 + N_2O + 5H_2O$

$Zn + 4HNO_{3(conc.)} \longrightarrow Zn(NO_3)_2 + 2NO_2 + 2H_2O$

27. (a)

Name	Formula	Oxidation state
Orthophosphorous acid	H_3PO_3	+3
Pyrophosphorous acid	$H_4P_2O_5$	+3
Hypophosphoric acid	$H_4P_2O_6$	+4
Pyrophosphoric acid	$H_4P_2O_7$	+5

28. (c) (a) $\overset{+6-1}{XeF_6} + \overset{+1-2}{H_2O} \longrightarrow \overset{+6-2-1}{XeOF_4} + \overset{+1-1}{2HF}$

No change in oxidation numbers, hence, no redox reaction occurs.

(b) $\overset{+6-1}{XeF_6} + \overset{+1-2}{2H_2O} \longrightarrow \overset{+6-2-1}{XeO_2F_2} + \overset{+1-1}{4HF}$

No change in oxidation numbers hence, no redox reaction occurs.

decrease in O.No. (reduction)

$$\overset{+4\,-1}{XeF_4} + \overset{+1\,-1}{O_2F_2} \longrightarrow \overset{+6\,-1}{XeF_6} + \overset{0}{O_2}$$

(c) increase in O.No. (oxidation)

Hence, it is a redox reaction.

(d) $\overset{+2-1}{XeF_2} + \overset{+5-1}{PF_5^-} \longrightarrow \overset{+2-1}{[XeF]^+}\ \overset{+5-1}{PF_6^-}$

No change in oxidation numbers hence, no redox reaction occurs.

29. (a) When Cl_2 gas reacts with cold and dilute aqueous NaOH, chloride (Cl^-) and hypochlorite (ClO^-) ions are formed.

$$\underset{}{Cl_2} + \underset{(dil.)}{2NaOH} \xrightarrow{Cold} \underset{\text{(Sodium chloride)}}{NaCl} + \underset{\text{(Sodium hypochloride)}}{NaOCl} + H_2O$$

30. (b) $NH_4NO_2 \xrightarrow{\Delta} N_2 + 2H_2O$

$(NH_4)_2SO_4 \xrightarrow{\Delta} NH_3 + H_2SO_4$

$Ba(N_3)_2 \xrightarrow{\Delta} Ba + 3N_2$

$(NH_4)_2Cr_2O_7 \xrightarrow{\Delta} N_2 + 4H_2O + Cr_2O_3$

31. (b) BCl_3 and $AlCl_3$ are e^- deficient and thus act Lewis acid.

32. (b) $B_2H_6 + 3O_2 \longrightarrow B_2O_3 + 3H_2O + \text{Heat}$

$B_2H_6 + 6H_2O \longrightarrow 2H_3BO_3 + 6H_2$

33. (b) (a) Pb_3O_4 is insoluble in water or do not react with water.

(b) $2KO_2 + 2H_2O \longrightarrow 2KOH + H_2O_2 + O_{2(g)} \uparrow$

(c) $Na_2O_2 + 2H_2O \longrightarrow 2NaOH + H_2O_2$

(d) $Li_2O_2 + 2H_2O \longrightarrow 2LiOH + H_2O_2$

34. (a) $H_3C\text{–}C(CH_3)_2\text{–}Br + NaOH \xrightarrow[\text{(First order)}]{S_N1} H_3C\text{–}C(CH_3)_2\text{–}OH + NaBr$

$t = 0$	P_0	0
t	P	$P_0 - P$

Rate = $k[(CH_3)_3C\text{–}Br]$ $\quad \ln \dfrac{P_0}{P} = kt$

$t_{1/2} = \dfrac{0.693}{k}$ $\quad \ln \dfrac{P}{P_0} = -kt$

$$\frac{[Q]}{[P]_0} = \frac{[P_0] - [P]}{[P_0]} = 1 - \frac{[P]}{[P_0]} = 1 - e^{-kt}$$

35. (a) $Au + HNO_3 + 4HCl \longrightarrow HAuCl_4 + NO + 2H_2O$

Gas evolved is nitric oxide (NO).

36. (b)

$$Pb(NO_3)_2(s) \xrightarrow{\Delta} PbO(s) + 2NO_2(g) + \frac{1}{2}O_2(g)$$
$$\quad (A)$$

$$2NO_2(g) \xrightarrow{\text{Cooling}} \underset{(B)}{N_2O_4(s)}$$

$$N_2O_4 + 2NO \longrightarrow \underset{(C)}{2N_2O_3}$$

Oxidation number of nitrogen in N_2O_3 is +3 .

37. (d)

$$P_4 + 3NaOH + 3H_2O \longrightarrow \underset{(X)}{3NaH_2PO_2} + PH_3$$

$$NaH_2PO_2 + HCl \longrightarrow NaCl + \underset{(Y)}{H_3PO_2}$$

H_3PO_2 is monobasic acid.

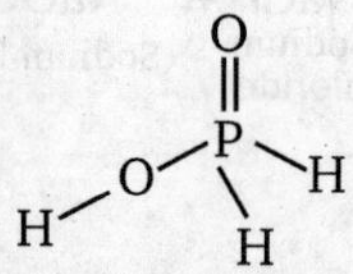

Basicity = 1

38. (a) Pyrophosphoric acid ($H_4P_2O_7$) structure is:

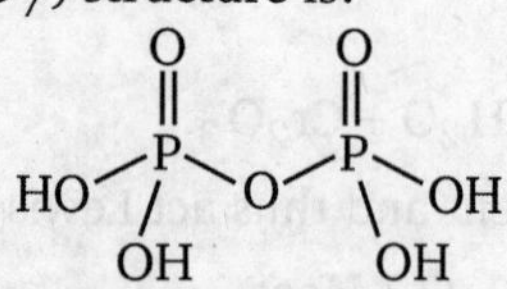

Bonds	Number
P — OH	4
P = O	2
P — O — P	1

39. (a) (A) $H_2O_2 \longrightarrow H_2O + \frac{1}{2}O_2(g)\uparrow$

(B) $2KClO_3 \longrightarrow 2KCl + 3O_2(g)\uparrow$

$2Pb(NO_3)_2 \longrightarrow 2PbO + 4NO_2 + O_2(g)\uparrow$

$2NaNO_3 \longrightarrow 2NaNO_3 + O_2(g)\uparrow$

(C)

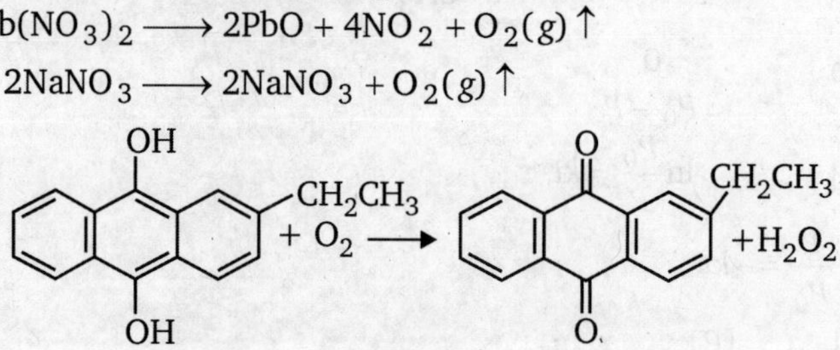

(D) Preparation of sodium perborate.

$$Na_2B_4O_7 + 2NaOH \longrightarrow 4NaBO_2 + H_2O$$

$$2NaBO_2 + 2H_2O_2 \longrightarrow \underset{\text{Sodium perborate}}{NaB_2O_4(OH)_4}$$

40. (d) Structure of $S_2O_8^{2-}$

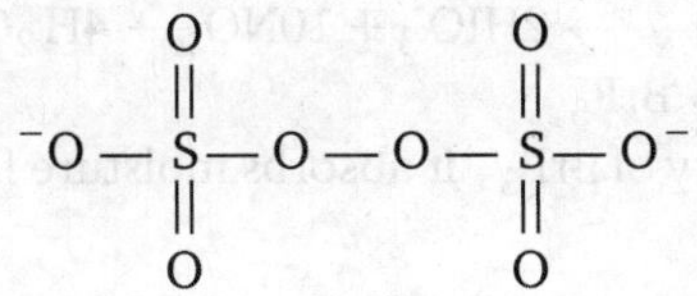

Structure of rhombic sulphur

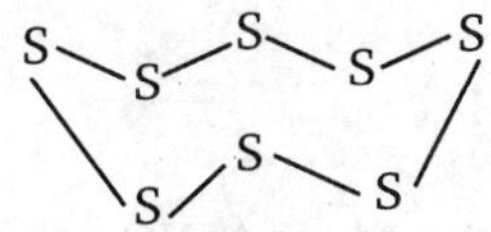

41. (a) $6NaOH + 3Cl_2 \longrightarrow \underset{(A)}{NaClO_3} + 5NaCl + 3H_2O$

$2Ca(OH)_2 + Cl_2 \longrightarrow \underset{(B)}{Ca(OCl)_2} + CaCl_2 + H_2O$

42. (a) $\underset{sp^3d^2}{\overline{Xe}F_4} + SbF_5 \longrightarrow \underset{sp^3d}{[\overline{Xe}F_3]^+}[SbF_6]^-$

43. (a)

Oxy acid	Formula	Oxidation state of 'P'
Hypophosphorous acid	H_3PO_2	+1
Orthophosphoric acid	H_3PO_4	+5
Hypophosphoric acid	$H_4P_2O_6$	+4
Orthophosphorous acid	H_3PO_3	+3

44. (c) $HO\overset{+1}{Cl} + H_2O_2 \longrightarrow H_3O^+ + \overset{-1}{Cl^-} + O_2$

Since, H_2O_2, reduces chlorine from +1 to 1, it acts as reducing agent.

❖ $\overset{0}{I_2} + H_2O_2 + OH^- \longrightarrow \overset{-1}{2I^-} + 2H_2O + O_2$

H_2O_2 reduces chlorine from 0 to –1, so it acts as reducing agent.

45. (d) Pyrophosphoric acid is $H_4P_2O_7$ with highest number of oxygen atoms.

46. (d) Red phosphorus is obtained by heating white phosphorus at 573 K in inert atmosphere. When red phosphorus is heated in a sealed tube at 803 K, we get α-black phosphorus. β-Black phosphorus is prepared by heating white phosphorus at 473 K under high pressure.

47. (a) Sulphur have a general tendency to form two double bond.

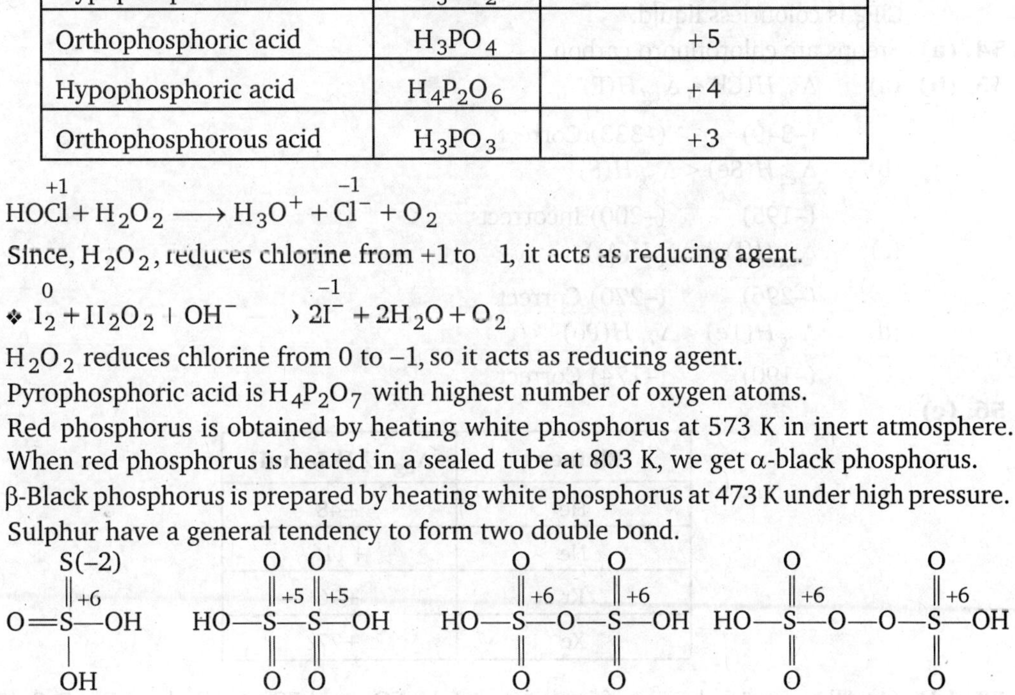

$H_2S_2O_3$ $\quad$ $H_2S_2O_6$ $\quad$ $H_2S_2O_7$ $\quad$ $H_2S_2O_8$

48. (c) The reaction involved is :

$$I_2 + 10HNO_3(\text{conc.}) \longrightarrow 2HIO_3 + 10NO_2 + 4H_2O$$

49. (b) $Br_2 + \text{Excess } F_2 \longrightarrow BrF_5$

[**Hint :** Due to high reactivity of BrF_5, it absorbs moisture from surrounding resulting into formation of $HBrO_3$]

50. (b)

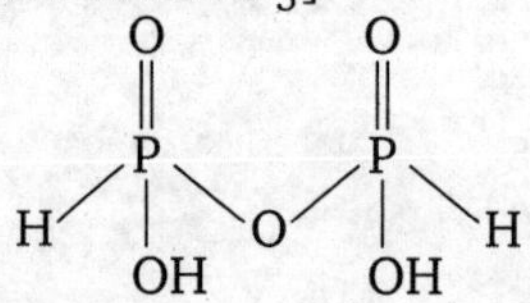

Oxyacid having P—H bond can reduce $AgNO_3$ to Ag.

51. (a) $P_4 + 8SOCl_2 \longrightarrow 4PCl_3 + 2S_2Cl_2 + 4SO_2$

52. (a) Bond dissociation energy of *E*—H bond in hydrides of group 16 follows the order :

$$H_2O > H_2S > H_2Se > H_2Te$$

53. (c)

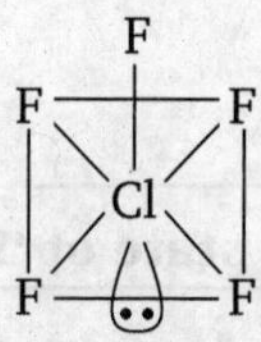

ClF_5 is colourless liquid.

54. (a) Freons are chlorofluoro carbon.

55. (b) (a) $\Delta_{eg}H(Cl) < \Delta_{eg}H(F)$

(−349) (−333) Correct

(b) $\Delta_{eg}H(Se) < \Delta_{eg}H(S)$

(−195) (−200) Incorrect

(c) $\Delta_{eg}H(I) < \Delta_{eg}H(At)$

(−296) (−270) Correct

(d) $\Delta_{eg}H(Te) < \Delta_{eg}H(Po)$

(−190) (−174) Correct

56. (c)

Element	$\Delta_{eg}H$ [kJ/mol]
He	+48
Ne	+116
Kr	+96
Xe	+77

57. (d) (A) All group 16 elements form oxides of the EO_2 and EO_3 type where E = S, Se, Te or Po.

(B) SO_2 is reducing while TeO_2 is an oxidising agent.

(C) The reducing property increases from H_2S to H_2Te down the group.

(D) 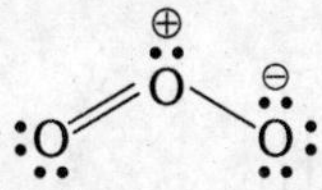 have six lone pairs.

58. (c) SiO_2 and GeO_2 are acidic and SnO, PbO are amphoteric.
Carbon does not have *d*-orbitals so can not form $p\pi$-$d\pi$ bond with itself. Due to properties of catenation and $p\pi$-$p\pi$ bond formation, carbon is able to show allotropic forms.

59. (d) $N - N < P - P$: single (σ) bond strength due to L.P.-L.P. replusion and maximum possible covalency of nitrogen is 4.

60. (d) SO_2 can oxidise as well as reduce.
Hence, it can act as both oxidising and reducing agent.

61. (d) Order of M.P. of group 14 : C > Si > Ge > Pb > Sn

element	M.P. (°C)
$Z = 6 = C$	3730
$Z = 14 = Si$	1410
$Z = 32 = Ge$	937
$Z = 50 = Sn$	232
$Z = 82 = Pb$	327

62. (b) Oxygen exists as O_2 (Atomicity = 2)
Sulphur exists as S_8 (Atomicity = 8)
Hence, melting point and boiling point of sulphur are significantly large compared to oxygen.

63. (c) Among Al, In, Tl, Ge, Sn, Pb, the metal having highest IE_1 is Ge and lowest IE_1 is In.
Most stable oxidation state of Ge is +4 and In is +3.

B. Objective Questions

[One or more than one option(s) is/are correct]

1. (b) $2PbO_2 + 4HNO_3 \longrightarrow 2Pb(NO_3)_2 + 2H_2O + O_2$

2. (b, c, d)

$$Ca(HCO_3)_2 + Ca(OH)_2 \longrightarrow 2CaCO_3 \downarrow + 2H_2O$$

(Clarke's method)

$$NaOCl + H_2O \rightleftharpoons HOCl + NaOH$$

$$OH^- + HCO_3^- \longrightarrow CO_3^{2-} + H_2O$$

$$Ca(HCO_3)_2 + Na_2CO_3 \longrightarrow CaCO_3 \downarrow + 2NaHCO_3$$

3. (b, d)

(a) H_3BO_3 is a weak monobasic Lewis acid and accepts OH^- from H_2O.

$$H_3BO_3 + H - OH \rightleftharpoons B(OH)_4^- + H^+ \quad ...(i)$$

(b) Equilibrium (i) is shifted in forward direction by the addition of syn-diols like ethylene glycol which forms a stable complex with $B(OH)_4^-$ and thus more H^+ ions are obtained.

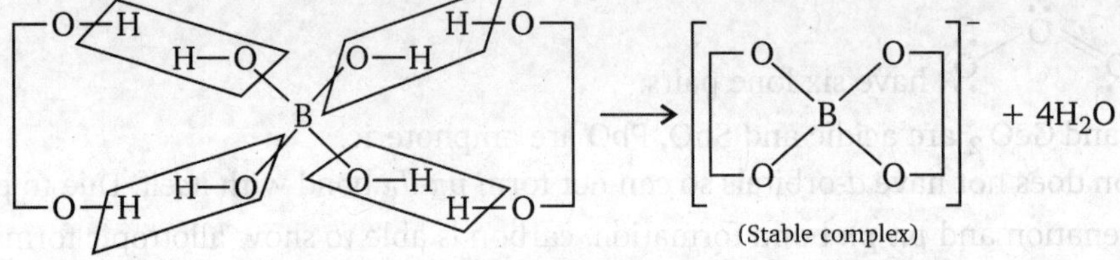

(c) It has a planar sheet like structure due to hydrogen bonding.

(d) H_3BO_3 is a weak electrolyte in water.

4. (a, d)

The formula of borax is $Na_2[B_4O_5(OH)_4]\cdot 8H_2O$ which contains the tetranuclear unit $[B_4O_5(OH)_4]^{2-}$.

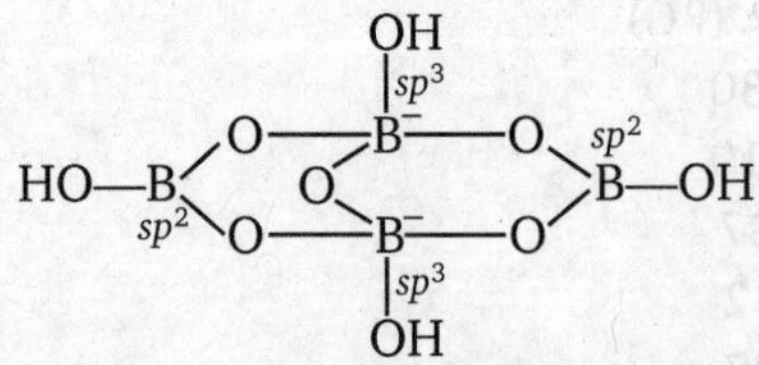

Only two B atoms lie in the same plane as two B atoms are sp^2 hybridized and other two B atoms are sp^3 hybridized.

5. (a, b, c) Both $Al(CH_3)_3$ and BH_3 has $3c$-$2e$ bonds in the dimeric structure.

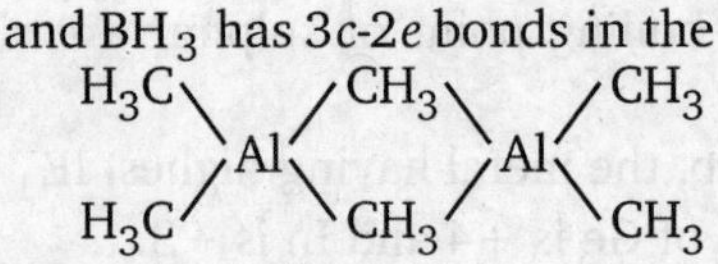

BCl_3 is stronger Lewis acid than $AlCl_3$.

6. (a, b or a, b, c)

(a) $SnCl_2.2H_2O$ is a reducing agent since Sn^{2+} tends to convert into Sn^{4+}.

(b) $\underset{\text{(Amphoteric)}}{SnO_2} + \underset{\text{(Base)}}{2KOH(aq.)} + 2H_2O \longrightarrow K_2[Sn(OH)_6]$

(c) First group cations (Pb^{2+}) form insoluble chloride with HCl that is $PbCl_2$, However it is slightly soluble in water and therefore lead +2 ion is never completely precipitated on adding hydrochloric acid in test sample of Pb^{2+}, rest of the Pb^{2+} ions are quantitatively precipitated with H_2S in acidic medium.

So that we can say that filtrate of first group contain solution of $PbCl_2$ in HCl which contains Pb^{2+} and Cl^-.

However in the presence of conc. HCl or excess HCl it can produce $H_2[PbCl_4]$.

So, we can conclude (a), (b) or (a), (b), (c) should be answers.

(d) $\underset{\substack{(2PbO.PbO_2)\\ \text{(Mixture of oxides)}}}{Pb_3O_4} + 4HNO_3 \longrightarrow PbO_2(\downarrow) + 2Pb(NO_3)_2 + 2H_2O$

It is not a redox reaction.

7. (a, b, c)

N_2O: $:N\equiv N\longrightarrow \ddot{\underset{\cdot\cdot}{O}}:$

N_2O_3: 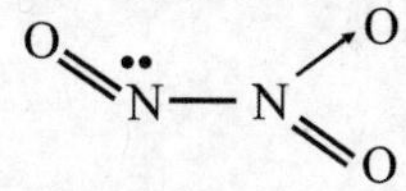

N_2O_4: 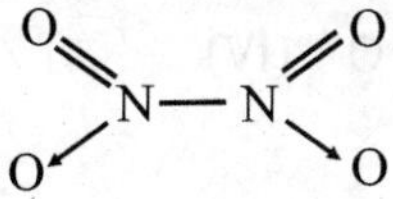

N_2O_5: 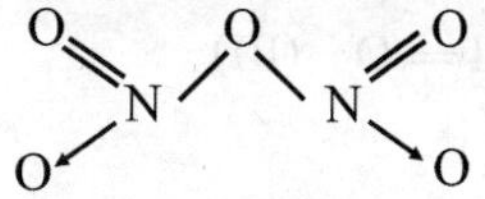

8. (b, d)

Diamond is the hardest substance known and C — C bond length is 1.54 Å in diamond. It is non- conductor of electricity. While in graphite, after sp^2 hybridisation one electron is free and it overlaps with another electron to form π-bond, thus bond length in graphite is shorter (1.42 Å) and bond order is higher than diamond. The π-electron is free to move thus graphite is good conductor of electricity but graphite is bad conductor of heat than diamond.

9. (a, c, d)

With $AgNO_3$, HCl, HBr and HI give precipitate.

$$AgNO_3 + HCl \longrightarrow \underset{\text{(white ppt.)}}{AgCl} + HNO_3$$

$$AgNO_3 + HBr \longrightarrow \underset{\text{(pale yellow ppt.)}}{AgBr} + HNO_3$$

$$AgNO_3 + HI \longrightarrow \underset{\text{(yellow ppt.)}}{AgI} + HNO_3$$

But HF does not give any precipitate, AgF is formed which is soluble in water.

$$AgBr + 2Na_2S_2O_3 \longrightarrow \underset{\substack{\text{Sodium argentothiosulphate} \\ \text{(colourless)}}}{Na_3[Ag(S_2O_3)_2]} + NaBr$$

Similar reactions are observed with AgCl and AgI.

10. (a, b, c)

(a) Sodium (Na) when dissolved in excess liquid ammonia, forms a blue coloured paramagnetic solution.

(b) $K + O_2 \longrightarrow \underset{\text{(Potassium superoxide)}}{KO_2}$ and KO_2 is paramagnetic.

(c) $3Cu + \underset{\text{(Dilute)}}{8HNO_3} \longrightarrow 3Cu(NO_3)_2 + 2NO + 4H_2O$

Where "NO" is paramagnetic.

(d) 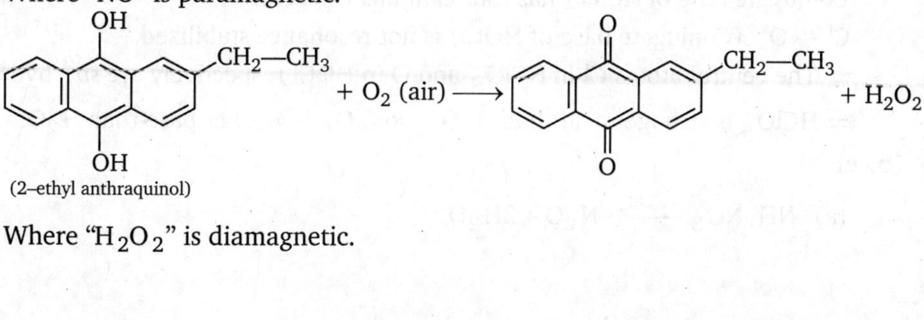

(2–ethyl anthraquinol)

Where "H_2O_2" is diamagnetic.

11. (b, c)

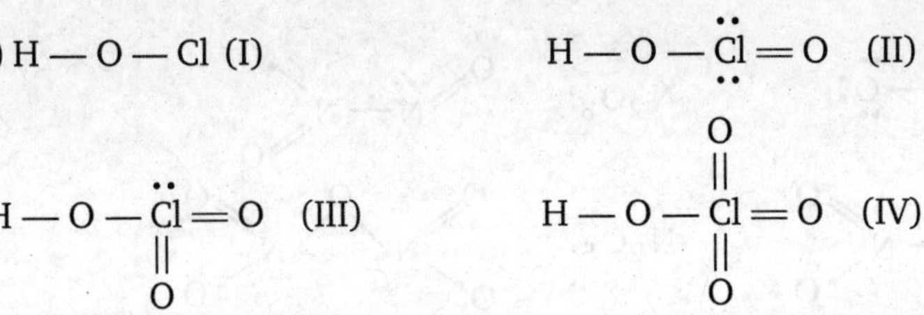

12. (b, c)

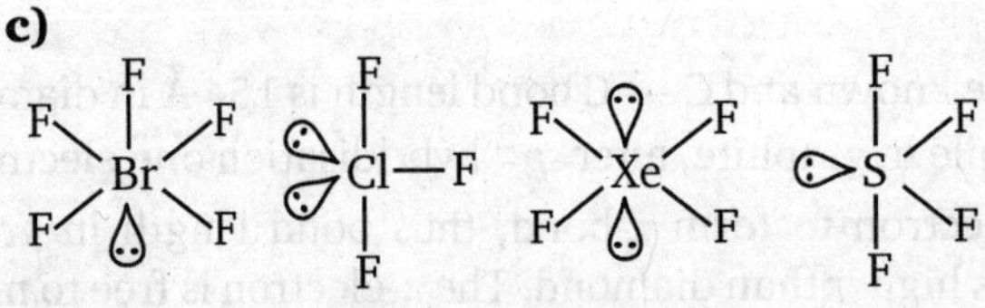

13. (b, d)

$$P_4O_{10} + 4HNO_3 \longrightarrow 2N_2O_5 + 4HPO_3$$

N_2O_5 cannot be obtained by reaction of P_4 and HNO_3,

$$P_4 + 20HNO_3 \longrightarrow 20NO_2 + 4H_3PO_4 + 4H_2O$$

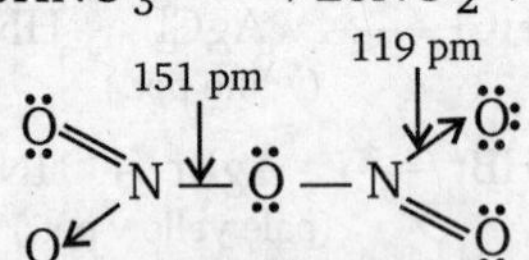

N_2O_5 (Planar)

Hence, it is diamagnetic and does not have N–N bond.

N_2O_5 is decomposed by alkali metals,

$$N_2O_5 + Na \longrightarrow NaNO_3 + \underset{\text{(Brown gas)}}{NO_2}$$

14. (b, c)

Highest occupied molecular orbital (HOMO) $\Rightarrow \pi^*$

Lowest unoccupied molecular orbital (LUMO) $\Rightarrow \sigma^*$

On descending the group gap between π^* and σ^* decreases.

15. (a, b, d)

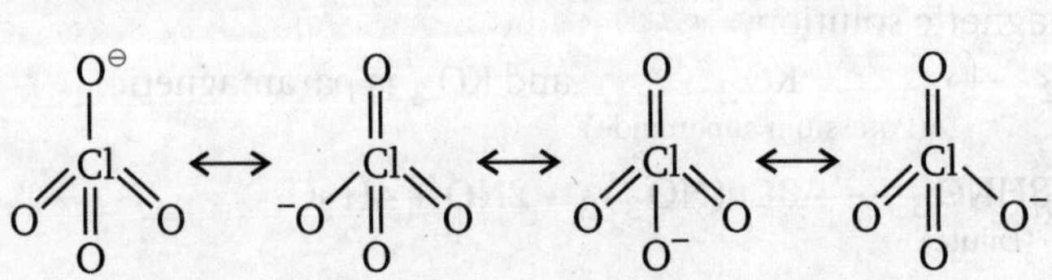

Conjugate base of $HClO_4$ has four canonical structures.

$Cl-O^{\ominus}$ (Conjugate base of HOCl) is not resonance stabilized.

$\Rightarrow$ The central atoms Cl in $HClO_4$ and O in HOCl respectively are sp^3 hybridized.

$\Rightarrow$ $HClO_4$ is stronger acid than H_3O^+, so ClO_4^- is weaker base than H_2O.

16. (b, c)

(a) $NH_4NO_3 \xrightarrow{300°C} N_2O + 2H_2O$

(b) $(NH_4)_2Cr_2O_7 \xrightarrow[300°C]{\Delta} N_2 + Cr_2O_3 + 4H_2O$

(c) $Ba(N_3)_2 \xrightarrow{\Delta} Ba + 3N_2$

(d) $\underset{\text{Ionic comp.}}{Mg_3N_2} \xrightarrow[300°C]{} \text{do not decompose}$

17. (a, b, c)

(a) Basic nature increase down the group in oxide basic strength $Bi_2O_5 > N_2O_5$.

(b) NF_3 is more covalent than BiF_3 according to Fajan's rule, smaller size of cation brings more covalent character.

(c) Due to H bond NH_3 has higher b.pt. than PH_3.

18. (a, b)

$$\underset{(Q)}{SnCl_2} + Cl^- \longrightarrow \underset{(X)}{SnCl_3^-}$$

X is sp^3 hybridised with one lone pair of electrons.

$$\underset{(Q)}{SnCl_2} + Me_3N \longrightarrow \underset{(Y)}{SnCl_2 \cdot [N(CH_3)_3]}$$

$SnCl_2$ is a Lewis acid while Me_3N acts as Lewis base hence, there is a coordinate bond between them.

$$\underset{(Q)}{SnCl_2} + 2CuCl_2 \longrightarrow \underset{(Z)}{SnCl_4} + 2CuCl$$

Sn in $SnCl_4$ has +4 oxidation state and $SnCl_4$ is tetrahedral with no lone pair.

19. (b, c, d)

$$Au + 4H^+ + NO_3^- + 4Cl^- \longrightarrow [AuCl_4]^{\ominus} + NO + 2H_2O$$

(b) NOCl/NO is formed.

(d) $[AuCl_4]^-$ gets formed; Au(+3).

20. (a, b, d) Hypochlorite ion : $ClO^{\ominus}$

Chloratc ion : $ClO_3^{\ominus}$

Perchlorate ion : $ClO_4^{\ominus}$

(a) Acidic order : $\overset{+1}{HClO} < \overset{+5}{HClO_3} < \overset{+7}{HClO_4}$

Conjugate base order : $ClO^- > ClO_3^- > ClO_4^-$

(b) Hypochlorite ion ($ClO^{\ominus}$): $:\ddot{\underset{\cdot\cdot}{Cl}}—\overset{\ominus}{\ddot{\underset{\cdot\cdot}{O}}}:$

(Linear shape)

Chlorate ion ($ClO_3^{\ominus}$) : 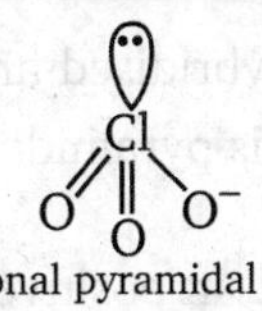

(Trigonal pyramidal shape)

Perchlorate ion ($ClO_4^{\ominus}$): 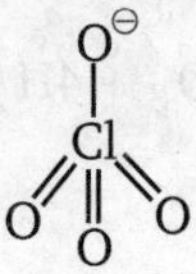

(Perfect tetrahedral shape due to resonance)

In chlorate ion bond angle changes due to presence of lone pair on chlorine atom. While hypochlorite ion is linear and perchlorate ion is tetrahedral and there is no effect of lone pair on hypochlorite ion.

(c) Disproportionation reaction of

(i) Hypochlorite ion : $3ClO^{\ominus} \longrightarrow 2Cl^{-} + ClO_3^{\ominus}$

(ii) Chlorate ion : $4ClO_3^{\ominus} \longrightarrow 3ClO_4^{\ominus} + Cl^{\ominus}$

(d) $ClO^{-} + SO_3^{2-} \longrightarrow SO_4^{2-} + Cl^{\ominus}$

C. Matching Type Problems

1. a — q, s, b — q, s, c—r, d — q, r

$$Bi^{3+} + H_2O \longrightarrow [BiO]^{+} + 2H^{+}$$

$$NaAlO_2 + 2H_2O \longrightarrow Al(OH)_3 + NaOH$$

$$2SiO_4^{4-} + 2H^{+} \longrightarrow Si_2O_7^{6-} + H_2O$$

$$Na_2B_4O_7 + 7H_2O \rightleftharpoons 4B(OH)_3 + 2NaOH$$

$$Na_2B_4O_7 + 2HCl + 5H_2O \longrightarrow 4B(OH)_3 + 2NaCl$$

2. a — p, s; b — q, s; c — r, t; d — q, t.

$$3Cu + \underset{\text{dil.}}{8HNO_3} \longrightarrow 3Cu(NO_3)_2 + 2NO + 4H_2O$$

$$Cu + \underset{\text{conc.}}{4HNO_3} \longrightarrow Cu(NO_3)_2 + 2NO_2 + 2H_2O$$

$$4Zn + \underset{\text{dil.}}{10HNO_3} \longrightarrow 4Zn(NO_3)_2 + N_2O + 5H_2O$$

$$4Zn + \underset{\text{conc.}}{4HNO_3} \longrightarrow Zn(NO_3)_2 + 2NO_2 + 2H_2O$$

D. Passage Based Problems

Passage-1

1. (a) Argon is used in arc welding to produce inert atmosphere as it is chemically inert gas.

2. (c) In XeO_3 central Xe atom is sp^3 hybridised and one sp^3 hybrid orbital is occupied by lone pair, thus shape is pyramidal.

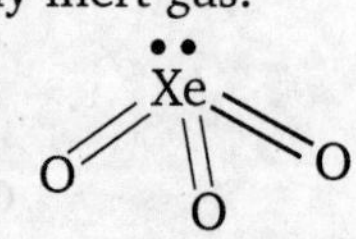

3. (a) Because Xe has complete octet and in positive oxidation states, it acts as oxidising agent. *e.g.*,

$$XeF_4 + 4I^- \longrightarrow Xe + 2I_2 + 4F^-$$

$$XeF_6 + 3H_2 \longrightarrow Xe + 6HF$$

Passage-2

1. (c) Nitrates are less abundant in earth crust, as they are water soluble and dissolves in water flowing over or leading through crust. They are attacked by microbes present in the soil also making them less abundant.

2. (c) NH_3 is more basic than PH_3 *i.e.,* it is better electron donor than PH_3.

3. (b) $P_4 + 3NaOH + 3H_2O \longrightarrow 3NaH_2\overset{+1}{P}O_2 + \overset{-3}{P}H_3$

(Disproportionation reaction)

Passage-3

1. (a) ; **2. (b)**

$$2KClO_3 \xrightarrow{MnO_2} 2KCl + \underset{W}{3O_2}$$

$$P_4 + 5O_2 \longrightarrow \underset{X}{P_4O_{10}}$$

$$P_4O_{10} + 4HNO_3 \longrightarrow 4HPO_3 + 2N_2O_5$$

E. Statement and Explanation Type Problems

1. (c) Si contains vacant *d*-orbitals so it accepts lone pair of H_2O and gets hydrolysed. On the other hand C in CCl_4 can't do so due to absence of vacant *d*-orbitals.

2. (a) Small, highly charged B^{3+} ion polarises anion to larger extent and so favours covalent bonding.

3. (c) In aqueous solution of boric acid $B(OH)_3$, it accepts OH^- and forms $[B(OH)_4]^-$. The proton is liberated by H_2O.

$$B(OH)_3 + HOH \rightleftharpoons [B(OH)_4]^- + H^+$$

4. (c) Pb^{4+} is less stable than Sn^{4+} and so better oxidant. But inert pair effect explains the poor stability of higher oxidation states for heavier members of gp. 14 elements. Thus Pb^{2+} is more stable than Pb^{4+}.

5. (b) Dinitrogen and dioxygen combine to form nitric oxide when the mixture is heated to 2273 – 3273 K in an electric arc.

6. (c) Order of I.E. (in kJ/mol) :

C	>	Si	>	Ge	>	Sn	<	Pb
1086		786		761		708		715

F. Integer Answer Type Problems

1. Co-ordination number of Al is 6. It is exists in c.c.p. lattice with 6 co-ordinate layer structure.

2. $Be_nAl_2Si_6O_{18}$ (Beryl)
(according to charge balance in a molecule)

$$2n + 6 + 24 - 36 = 0$$
$$n = 3$$

3. (4) $PCl_5 + SO_2 \longrightarrow POCl_3 + SOCl_2$ (Thionyl chloride)

$$6PCl_5 + P_4O_{10} \longrightarrow 10POCl_3$$

When equimolar amounts of PCl_5 and H_2O are used, the reaction is gentle :

$$PCl_5 + H_2O \longrightarrow POCl_3 + 2HCl$$

[**Note :** PCl_5 reacts violently with H_2O on complete hydrolysis to producę phosphoric acid.

$$PCl_5 + 4H_2O \longrightarrow H_3PO_4 + 5HCl \,]$$

$$2PCl_5 + H_2SO_4 \longrightarrow 2POCl_3 + \underset{\text{(sulphuryl chloride)}}{SO_2Cl_2} + 2HCl$$

4. (6) $B_2H_6 + 6MeOH \longrightarrow 2B(OMe)_3 + 6H_2$

1 mole of B_2H_6 reacts with 6 mole of MeOH to give 2 moles of $B(OMe)_3$.

3 mole of B_2H_6 will react with 18 mole of MeOH to give 6 moles of $B(OMe)_3$.

5. (19) $XeF_4 + O_2F_2 \xrightarrow{143\text{ K}} \underset{(Y)}{XeF_6} + O_2$

In structure of XeF_6, one lone pair on Xe atom, three lone pairs on each F atom are present.

Thus total 19 lone pairs of electrons are present on molecule Y.

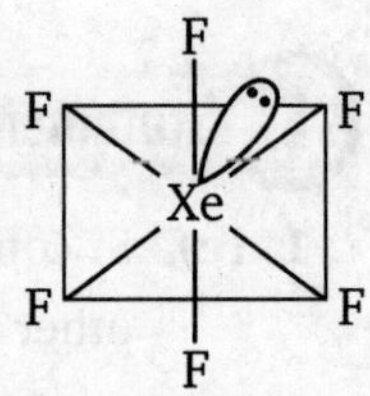

6. (1.67) $3Cl_2 + \underset{\text{Hot \& conc.}}{6NaOH} \longrightarrow \underset{(X)}{5NaCl} + \underset{(Y)}{NaClO_3} + 3H_2O$

$$\underset{(X)}{NaCl} + AgNO_3 \longrightarrow \underset{\text{(white ppt.)}}{AgCl} + NaNO_3$$

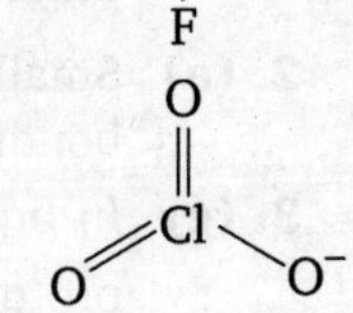

Average bond order between Cl and O atom in $NaClO_3 = \frac{5}{3} = 1.67$

7. (3) The number of halogen forming halic (V) acid is 3. $HClO_3$, $HBrO_3$ and HIO_3 are those acids.

8. (4) $P_4(s) + 3OH^-(aq.) + 3H_2O(l) \longrightarrow PH_3(g) + \underset{(A)}{3H_2PO_2^-(s)}$

$$\underset{\text{1 mole}}{\overset{+1}{H_2PO_2^-}} + \underset{\text{(Excess)}}{Ag^+} + H_2O \longrightarrow \overset{0}{Ag} + \overset{+5}{H_3PO_4} + H^+$$

$$[e^- + Ag^+ \longrightarrow Ag] \times 4$$
$$P^{+1} \longrightarrow P^{+5} + 4e^-$$
$$4Ag^+ + P \longrightarrow 4Ag + P^{+5}$$

9. (2) $C_2H_5OH + PCl_3 \longrightarrow C_2H_5Cl + \underset{(A)}{H_3PO_3}$

$$\underset{(A)}{H_3PO_3} + PCl_3 \longrightarrow \underset{(B)}{H_4P_2O_5} + HCL$$

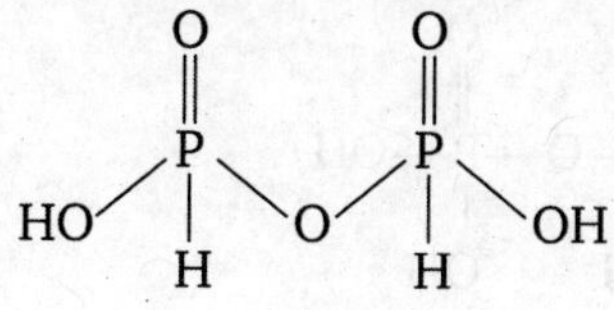

(*B*)

Number of non-ionisable protons in 'B' are 2.

10. (3) AB_5 species have square pyramidal structures. Hence, BrF_5, IF_5 and ClF_5.

11. (6) 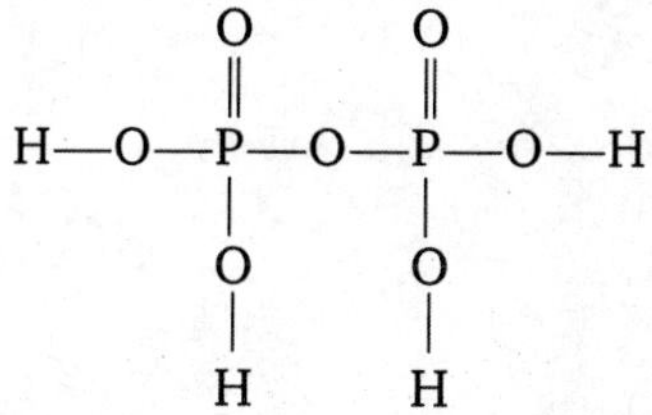

No. of σ bond = 12, No. of π bond = 2, $\frac{\sigma}{\pi} = \frac{12}{2} = 6$

So, Answer is 6.

12. (6)

N_2O_3 ⇒ 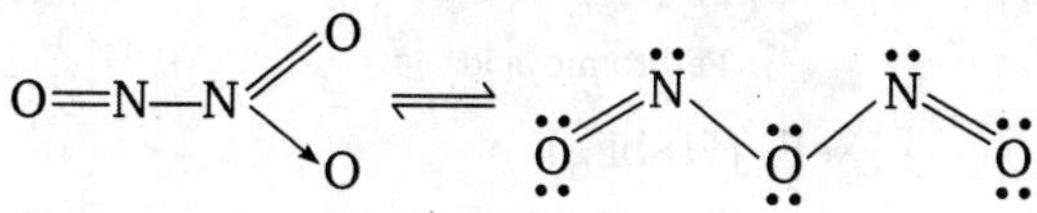

N_2O_5 ⇒ 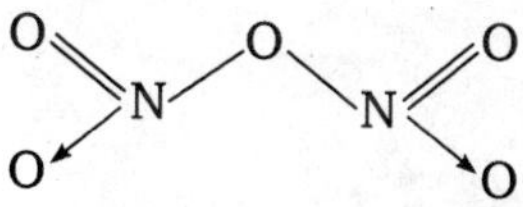

P_4O_6 ⇒ 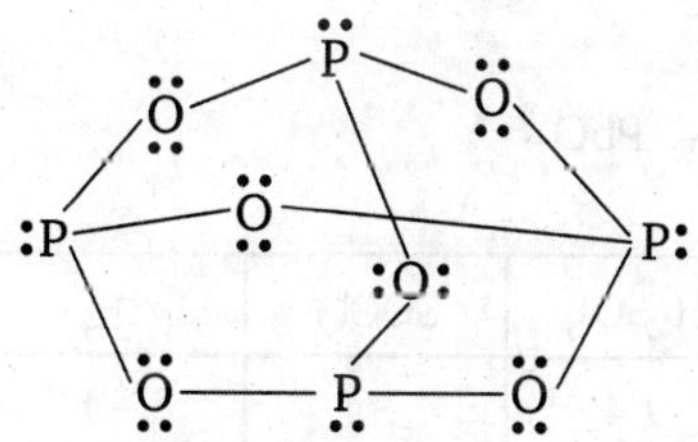

P_4O_7 ⇒ 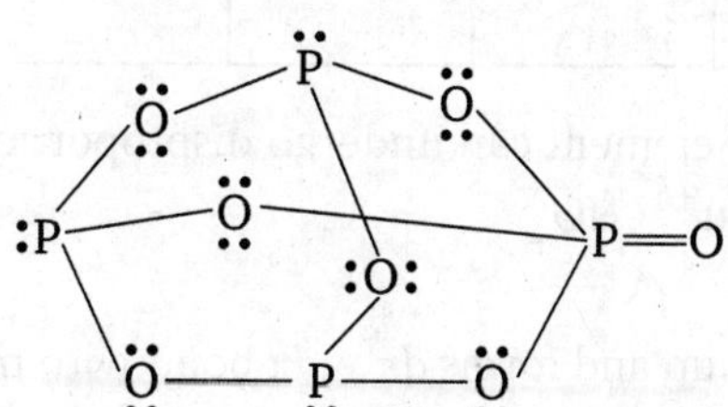

$H_4P_2O_5$ ⇒ 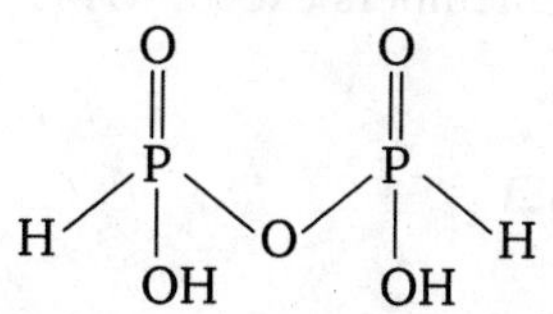

$H_5P_3O_{10}$ ⇒ 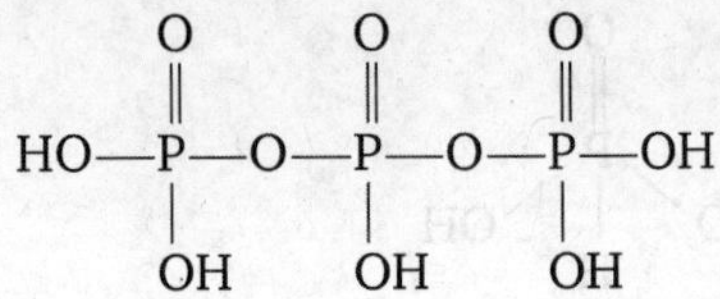

$H_2S_2O_3$ ⇒ 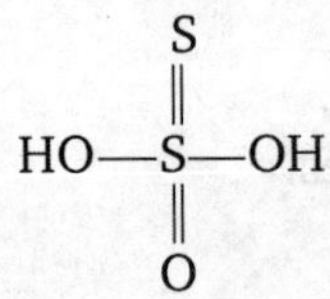

$H_2S_2O_5$ ⇒ 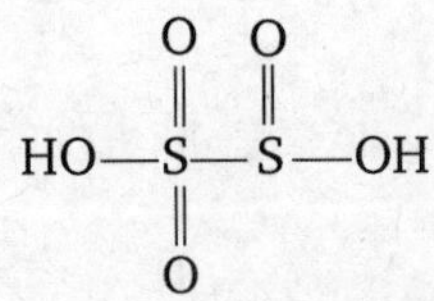

13. (1) IF_5 has 1 lone pair and IF_7 has 0 lone pair

∴ $1 + 0 = 1$

14. (12)

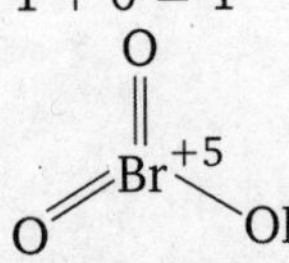

Bromic acid

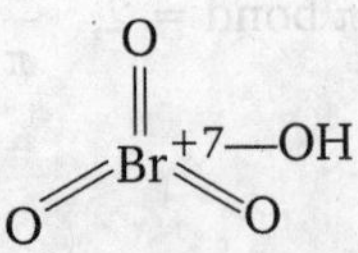

Perbromic acid

15. (11) $XeF_4 + SbF_5 \longrightarrow [XeF_3]^+[SbF_6]^-$

$m = 3, n = 1, y = 6, z = 1$

$m + n + y + z = 11$

16. (3) Acidic oxide : Cl_2O_7, SiO_2, N_2O_5

Neutral oxide : CO, NO, N_2O

Amphoteric oxide : Al_2O_3, SnO_2, PbO_2

17. (3)

Compounds	SO_3	H_2SO_3	$SOCl_2$	SF_4	$BaSO_4$	$H_2S_2O_7$
O.S. of Sulphur	+6	+4	+4	+4	+6	+6

18. (6) Intermediate oxidation state of element can undergo disproportionation.

$H_2O_2, ClO_3^-, P_4, Cl_2, Cu^{+1}, NO_2$

19. (15)

Phosphorus belongs to 15th group and forms $d\pi - d\pi$ bond with transition metal and PH_3 is strongest base among the other group members except NH_3.

❑❑❑

CHAPTER

7

d- & f-block Elements

A. Objective Questions

[Only one option is correct]

1. Which of the following arrangements does not represent the correct order of the property stated against it? **[JEE (Mains) 2013]**

(a) $V^{2+} < Cr^{2+} < Mn^{2+} < Fe^{2+}$: paramagnetic behaviour

(b) $Ni^{2+} < Co^{2+} < Fe^{2+} < Mn^{2+}$: ionic size

(c) $Co^{3+} < Fe^{3+} < Cr^{3+} < Sc^{3+}$: stability in aqueous solution

(d) $Sc < Ti < Cr < Mn$: number of oxidation states

2. Consider the following reaction :

$$x MnO_4^- + y C_2O_4^{2-} + z H^+ \longrightarrow x Mn^{2+} + 2y CO_2 + \frac{z}{2} H_2O$$

The values of x, y and z in the reaction are, respectively : **[JEE (Mains) 2013]**

(a) 5, 2 and 16 (b) 2, 5 and 8 (c) 2, 5 and 16 (d) 5, 2 and 8

3. Fou successive members of the first row transition elements are listed below with atomic numbers. Which one of them is expected to have the highest $E^\circ_{M^{3+}/M^{2+}}$ value? **[JEE (Mains) 2013]**

(a) $Cr(Z = 24)$ (b) $Mn(Z = 25)$ (c) $Fe(Z = 26)$ (d) $Co(Z = 27)$

4. Which series of reactions correctly represents chemical reactions related to iron and its compound ? **[JEE (Mains) 2014]**

(a) $Fe \xrightarrow{\text{dil.} H_2SO_4} FeSO_4 \xrightarrow{H_2SO_4, O_2} Fe_2(SO_4)_3 \xrightarrow{\text{Heat}} Fe$

(b) $Fe \xrightarrow{O_2, \text{heat}} FeO \xrightarrow{\text{dil.} H_2SO_4} FeSO_4 \xrightarrow{\text{Heat}} Fe$

(c) $Fe \xrightarrow{Cl_2, \text{heat}} FeCl_3 \xrightarrow{\text{heat, air}} FeCl_2 \xrightarrow{Zn} Fe$

(d) $Fe \xrightarrow{O_2, \text{heat}} Fe_3O_4 \xrightarrow{CO,\ 600°C} FeO \xrightarrow{CO,\ 700°C} Fe$

5. Which of the following compounds is metallic and ferromagnetic ? **[JEE (Mains) 2016]**

(a) TiO_2 (b) CrO_2 (c) VO_2 (d) MnO_2

6. The lanthanide ion that would show colour is : **[JEE (Mains) 2019]**
(a) Lu^{3+} (b) Sm^{3+} (c) La^{3+} (d) Gd^{3+}

7. The statement that is incorrect about the interstitial compound is : **[JEE (Mains) 2019]**
(a) they are chemically reactive (b) they have metallic conductivity
(c) they are very hard (d) they have high melting points

8. The incorrect statement is: **[JEE Main (Sept.) 2020]**
(a) In manganate and permanganate ions, the π-bonding takes place by overlap of *p*-orbitals of oxygen and *d*-orbitals of manganese.
(b) Manganate ion is green in colour and permanganate ion is purple in colour.
(c) Manganate and permanganate ions are paramagnetic.
(d) Manganate and permanganate ions are tetrahedral.

9. The molecular geometry of SF_6 is octahedral. What is the geometry of SF_4 (including lone pair(s) of electrons, if (any)? **[JEE Main (Sept.) 2020]**
(a) Tetrahedral (b) Trigonal bipyramidal
(c) Square planar (d) Pyramidal

10. For the following Assertion and Reason, the correct option is : **[JEE Main (Jan.) 2020]**
Assertion: For hydrogenation reactions, the catalytic activity increases from Group 5 to Group 11 metals with maximum activity shown by Group 7-9 elements.
Reason: The reactants are most strongly adsorbed on group 7-9 elements. Both assertion and reason are true and the reason is the correct.
(a) Assertion is not true, but reason is true.
(b) Both assertion and reason are false.
(c) Both assertion and reason are true, but the reason is not the correct explanation for the assertion.
(d) The assertion is true, but the reason is false.

11. '*X*' melts at low temperature and is a bad conductor of electricity in both liquid and solid state. *X* is: **[JEE Main (Jan.) 2020]**
(a) Silicon carbide (b) Mercury (c) Zinc sulphide (d) Carbon tetrachloride

12. The incorrect statement(s) among (A)-(C) is (are): **[JEE Main (Sept.) 2020]**
(A) W(VI) is more stable than Cr(VI).
(B) In the presence of HCl, permanganate titrations provide satisfactory results.
(C) Some lanthanoid oxides can be used as phosphors.
(a) (B) and (C) only (b) (A) only (c) (B) only (d) (A) and (B) only

13. The electronic configurations of bivalent europium and trivalent cerium are: (atomic number: Xe = 54, Ce = 58, Eu = 63) **[JEE Main (Jan.) 2020]**
(a) $[Xe]4f^7 6s^2$ and $[Xe]\ 4f^2 6s^2$ (b) $[Xe]4f^7 6s^2$ and $[Xe]\ 4f^1$
(c) $[Xe]4f^2$ and $[Xe]\ 4f^7$ (d) $[Xe]4f^4$ and $[Xe]\ 4f^9$

14. The correct electronic configuration and spin-only magnetic moment (BM) of Gd^{3+} ($Z = 64$), respectively, are: **[JEE Main (Sept.) 2020]**
(a) $[Xe]4f^7$ and 7.9 (b) $[Xe]4f^7$ and 8.9
(c) $[Xe]5f^7$ and 8.9 (d) $[Xe]5f^7$ and 7.9

15. The lanthanoid that does not show +4 oxidation state is: **[JEE Main (Sept.) 2020]**
(a) Dy (b) Ce (c) Eu (d) Tb

16. Given below are two statements:
Statement I : CeO_2 can be used for oxidation of aldehydes and ketones.
Statement II : Aqueous solution of $EuSO_4$ is a strong reducing agent.
In the light of the above statements, choose the correct answer from the options given below:
[JEE Main (Feb.) 2021]
(a) Statement I is false but statement II is true.
(b) Statement I is true but statement II is false.
(c) Both statement I and statement II are true.
(d) Both statement I and statement II are false.

17. Which one of the following lanthanoids does not form MO_2 ? [M is lanthanoid metal]
[JEE Main (Feb.) 2021]
(a) Pr (b) Dy (c) Nd (d) Yb

18. Given below are two statement : one is labelled as Assertion A and the other is labelled as Reason R :
Assertion A : Size of Bk^{3+} ion is less than Np^{3+} ion.
Reason R : The above is a consequence of the lanthanoid contraction.
In the light of the above statements, choose the correct answer from the options given below :
[JEE Main (March) 2021]
(a) A is false but R is true.
(b) Both A and R are true but R is not the correct explanation of A.
(c) Both A and R are true and R is the correct explanation of A.
(d) A is true but R is false.

19. Given below are two statements :
Statement I : The $E°$ value of Ce^{4+} / Ce^{3+} is +1.74 V.
Statement II : Ce is more stable in Ce^{4+} state than Ce^{3+} state.
In the light of the above statements, choose the most appropriate answer from the options given below : **[JEE Main (March) 2021]**
(a) Both statement I and statement II are correct.
(b) Statement I is incorrect but statement II is correct.
(c) Both statement I and statement II are incorrect.
(d) Statement I is correct but statement II is incorrect.

20. The electrode potential of M^{2+}/M of 3d-series elements shows positive value of :
[JEE Main (Feb.) 2021]
(a) Zn (b) Fe (c) Co (d) Cu

21. The incorrect statement among the following is: **[JEE Main (Feb.) 2021]**
(a) $VOSO_4$ is a reducing agent
(b) Cr_2O_3 is an amphoteric oxide
(c) RuO_4 is an oxidizing agent
(d) Red colour of ruby is due to the presence of Co^{3+}

22. Given below are two statements:

Statement I: Potassium permanganate on heating at 573K forms potassium manganate.

Statement II: Both potassium permanganate and potassium manganate are tetrahedral and paramagnetic in nature.

In the light of the above statements, choose the most appropriate answer from the options given below: **[JEE Main (March) 2021]**

(a) Statement I is true but statement II is false.
(b) Both statement I and statement II are true.
(c) Statement I is false but statement II is true.
(d) Both statement I and statement II are false.

23. Fex_2 and Fey_3 are known when x and y are : **[JEE Main (March) 2021]**

(a) $x =$ F, Cl, Br, I and $y =$ F, Cl, Br
(b) $x =$ F, Cl, Br and $y =$ F, Cl, Br, I
(c) $x =$ Cl, Br, I and $y =$ F, Cl, Br, I
(d) $x =$ F, Cl, Br, I and $y =$ F, Cl, Br, I

24. Match List-I with List-II. **[JEE Main (March) 2021]**

List-I		List-II	
(A)	Chlorophyll	(i)	Ruthenium
(B)	Vitamin-B_{12}	(ii)	Platinum
(C)	Anticancer drug	(iii)	Cobalt
(D)	Grubbs catalyst	(iv)	Magnesium

Choose the most appropriate answer from the options given below:

(a) A-iii, B-ii, C-iv, D-i
(b) A-iv, B-iii, C-ii, D-i
(c) A-iv, B-iii, C-i, D-ii
(d) A-iv, B-ii, C-iii, D-i

25. In following pairs, the one in which both transition metal ions are colourless is : **[JEE Main (July) 2022 (I)]**

(a) Sc^{3+}, Zn^{2+}
(b) Ti^{4+}, Cu^{2+}
(c) V^{2+}, Ti^{2+}
(d) Zn^{2+}, Mn^{2+}

26. The total number of Mn=O bonds in Mn_2O_7 is _____. **[JEE Main (July) 2022 (I)]**

(a) 4
(b) 5
(c) 6
(d) 3

27. Dihydrogen reacts with CuO to give : **[JEE Main (June) 2022 (I)]**

(a) CuH_2
(b) Cu
(c) Cu_2O
(d) $Cu(OH)_2$

28. The 'f' orbitals are half and completely filled, respectively in lanthanide ions (Given : Atomic no. Eu : 63 ; Sm : 62; Tm : 69; Tb : 65; Yb : 70; Dy : 66) **[JEE Main (June) 2022 (II)]**

(a) Eu^{2+} and Tm^{2+}
(b) Sm^{2+} and Tm^{3+}
(c) Tb^{4+} and Yb^{2+}
(d) Dy^{3+} and Yb^{3+}

29. In chromyl chloride, the number of d-electrons present on chromium is same as in : (Given at. no. of Ti : 22, V : 23, Cr : 24, Mn : 25, Fe : 26) **[JEE Main (April) 2023 (I)]**

(a) Ti(III)
(b) Fe(III)
(c) V(IV)
(d) Mn(VII)

30. The set of correct statements is : **[JEE Main (Jan.) 2023 (II)]**

(i) Manganese exhibits +7 oxidation state in its oxide.
(ii) Ruthenium and Osmium exhibit +8 oxidation in their oxides.

(iii) Sc shows +4 oxidation state which is oxidizing in nature.
(iv) Cr shows oxidising nature in +6 oxidation state.
(a) (ii) and (iii) (b) (i), (ii) and (iv)
(c) (i) and (iii) (d) (ii), (iii) and (iv)

31. Prolonged heating is avoided during the preparation of ferrous ammonium sulphate to : **[JEE Main (April) 2023 (I)]**
(a) prevent oxidation (b) prevent reduction
(c) prevent hydrolysis (d) prevent breaking

32. The correct order of basicity of oxides of vanadium is : **[JEE Main (Jan.) 2023 (I)]**
(a) $V_2O_3 > V_2O_4 > V_2O_5$ (b) $V_2O_3 > V_2O_5 > V_2O_4$
(c) $V_2O_5 > V_2O_4 > V_2O_3$ (d) $V_2O_4 > V_2O_3 > V_2O_5$

33. The pair of lanthanides in which both elements have high third-ionization energy is : **[JEE Main (April) 2023 (I)]**
(a) Eu, Gd (b) Eu, Yb (c) Lu, Yb (d) Dy, Gd

34. Strong reducing and oxidising agents among the following, respectively, are : **[JEE Main (April) 2023 (I)]**
(a) Ce^{4+} and Eu^{2+} (b) Ce^{4+} and Tb^{4+} (c) Ce^{3+} and Ce^{4+} (d) Eu^{2+} and Ce^{4+}

35. Given below are two statements :
Statement (I) : In the Lanthanoids, the formation of Ce^{4+} is favoured by its noble gas configuration.
Statement (II) : Ce^{4+} is a strong oxidant reverting to the common +3 state.
In the light of the above statements, choose the most appropriate answer from the options given below : **[JEE Main (Jan.) 2024]**
(a) Statement I is false but Statement II is true
(b) Both Statement I and Statement II are true
(c) Statement I is true but Statement II is false
(d) Both Statement I and Statement II are false

36. Choose the correct option having all the elements with d^{10} electronic configuration from the following : **[JEE Main (Jan.) 2024]**
(a) ^{27}Co, ^{28}Ni, ^{26}Fe, ^{24}Cr (b) ^{29}Cu, ^{30}Zn, ^{48}Cd, ^{47}Ag
(c) ^{46}Pd, ^{28}Ni, ^{26}Fe, ^{24}Cr (d) ^{28}Ni, ^{24}Cr, ^{26}Fe, ^{29}Cu

37. Which of the following statements are correct about Zn, Cd and Hg?
A. They exhibit high enthalpy of atomization as the *d*-subshell is full.
B. Zn and Cd do not show variable oxidation state while Hg shows +I and +II.
C. Compounds of Zn, Cd and Hg are paramagnetic in nature.
D. Zn, Cd and Hg are called soft metals.
Choose the most appropriate from the options given below : **[JEE Main (Jan.) 2024]**
(a) B, D only (b) B, C only (c) A, D only (d) C, D only

38. Diamagnetic Lanthanoid ions are : **[JEE Main (Jan.) 2024]**
(a) Nd^{3+} and Eu^{3+} (b) La^{3+} and Ce^{4+} (c) Nd^{3+} and Ce^{4+} (d) Lu^{3+} and Eu^{3+}

39. The correct IUPAC name of K_2MnO_4 is : **[JEE Main (Jan.) 2024]**
(a) Potassium tetraoxopermanganate (VI)
(b) Potassium tetraoxidomanganate (VI)
(c) Dipotassium tetraoxidomanganate (VII)
(d) Potassium tetraoxidomanganese (VI)

40. Lanthanoid ions with $4f^7$ configuration are : **[JEE Main (Jan.) 2025]**
(A) Eu^{2+} (B) Gd^{3+} (C) Eu^{3+} (D) Tb^{3+}
(E) Sm^{2+}
Choose the correct answer from the options given below :
(a) (A) and (B) only (b) (A) and (D) only
(c) (B) and (E) only (d) (B) and (C) only

41. Consider the following reactions

$$K_2Cr_2O_7 \xrightarrow[-H_2O]{KOH} [A] \xrightarrow[-H_2O]{H_2SO_4} [B] + K_2SO_4$$

The products $[A]$ and $[B]$, respectively are : **[JEE Main (Jan.) 2025]**
(a) $K_2Cr(OH)_6$ and Cr_2O_3 (b) K_2CrO_4 and Cr_2O_3
(c) K_2CrO_4 and $K_2Cr_2O_7$ (d) K_2CrO_4 and CrO

42. Which of the following ions is the strongest oxidizing agent?
(Atomic Number of Ce = 58, Eu = 63, Tb = 65, Lu = 71) **[JEE Main (Jan.) 2025]**
(a) Lu^{3+} (b) Eu^{2+}
(c) Tb^{4+} (d) Ce^{3+}

43. The amphoteric oxide among V_2O_3, V_2O_4 and V_2O_5 upon reaction with alkali leads to formation of an oxide anion. The oxidation state of V in the oxide anion is : **[JEE Main (Jan.) 2025]**
(a) +3 (b) +7 (c) +5 (d) +4

44. The correct option with order of melting points of the pairs (Mn, Fe), (Tc, Ru) and (Re, Os) is : **[JEE Main (Jan.) 2025]**
(a) Fe < Mn, Ru < Tc and Re < Os (b) Mn < Fe, Tc < Ru and Re < Os
(c) Mn < Fe, Tc < Ru and Os < Re (d) Fe < Mn, Ru < Tc and Os < Re

B. Objective Questions

[One or more than one option(s) is/are correct]

1. Reduction of the metal centre in aqueous permanganate ion involves : **(IIT 2011)**
(a) 3 electrons in neutral medium (b) 5 electrons in neutral medium
(c) 3 electrons in alkaline medium (d) 5 electrons in acidic medium

2. The correct statement(s) about Cr^{2+} and Mn^{3+} is(are): **[JEE (Advanced) 2015]**
[Atomic numbers of Cr = 24 and Mn = 25]
(a) Cr^{2+} is a reducing agent
(b) Mn^{3+} is an oxidizing agent

(c) Both Cr^{2+} and Mn^{3+} exhibit d^4 electronic configuration

(d) When Cr^{2+} is used as a reducing agent, the chromium ion attains d^5 electronic configuration

3. Fe^{3+} is reduced to Fe^{2+} by using : **[JEE (Advanced) 2015]**

(a) H_2O_2 in presence of NaOH
(b) Na_2O_2 in water
(c) H_2O_2 in presence of H_2SO_4
(d) Na_2O_2 in presence of H_2SO_4

4. Fusion of MnO_2 with KOH in presence of O_2 produces a salt W. Alkaline solution of W upon electrolytic oxidation yields another salt X. The manganese containing ions present in W and X, respectively, are Y and Z. Correct statement(s) is (are) : **[JEE (Advanced) 2019]**

(a) in both Y and Z, π-bonding occurs between p-orbitals of oxygen and d-orbitals of manganese

(b) both Y and Z are coloured and have tetrahedral shape

(c) in aqueous acidic solution, Y undergoes disproportionation reaction to give Z and MnO_2

(d) Y is diamagnetic in nature while Z is paramagnetic.

5. Consider the following reactions (unbalanced)

$$Zn + \text{hot conc. } H_2SO_4 \longrightarrow G + R + X$$

$$Zn + \text{conc. NaOH} \longrightarrow T + Q$$

$$G + H_2S + NH_4OH \longrightarrow Z \text{ (a precipitate)} + X + Y$$

Choose the correct option(s). **[JEE (Advanced) 2019]**

(a) Bond order of Q is 1 in its ground state.

(b) The oxidation state of Zn in T is +1.

(c) R is a V-shaped molecule.

(d) Z is dirty white in colour.

C. Passage Based Problems

Passage-1

When a metal rod **M** is dipped into an aqueous colourless concentrated solution of compound **N**, the solution turns light blue. Addition of aqueous NaCl to the blue solution gives a white precipitate **O**. Addition of aqueous NH_3 dissolves **O** and gives an intense blue solution.

(IIT 2011)

1. The metal rod **M** is :

(a) Fe (b) Cu (c) Ni (d) Co

2. The compound **N** is :

(a) $AgNO_3$ (b) $Zn(NO_3)_2$ (c) $Al(NO_3)_3$ (d) $Pb(NO_3)_2$

3. The final solution contains :

(a) $[Pb(NH_3)_4]^{2+}$ and $[CoCl_4]^{2-}$
(b) $[Al(NH_3)_4]^{3+}$ and $[Cu(NH_3)_4]^{2+}$
(c) $[Ag(NH_3)_2]^+$ and $[Cu(NH_3)_4]^{2+}$
(d) $[Ag(NH_3)_2]^+$ and $[Ni(NH_3)_6]^{2+}$

D. Integer Type Problems

1. The number of water molecule(s) directly bonded to the metal centre in $CuSO_4.5H_2O$ is: **(IIT 2009)**

2. An acidified solution of potassium chromate was layered with an equal volume of amyl alcohol. When it was shaken after the addition of 1 mL of 3% H_2O_2, a blue alcohol layer was obtained. The blue colour is due to the formation of a chromium (VI) compound 'X'. What is the number of oxygen atoms bonded to chromium through only single bonds in a molecule of X? **[JEE (Advanced) 2020]**

3. The oxidation states of transition metal atoms in $K_2Cr_2O_7$, $KMnO_4$ and K_2FeO_4, respectively, are x, y and z. The sum of x, y and z is ______. **[JEE Main (Sept.) 2020]**

4. Dichromate ion is treated with base, the oxidation number of Cr in the product formed is ______. **[JEE Main (Feb.) 2021]**

5. The number of statement(s) correct from the following for copper (at. no. 29) is/are ______.
   (A) Cu (II) complexes are always paramagnetic
   (B) Cu (I) complexes are generally colourless
   (C) Cu (I) is easily oxidized
   (D) In Fehling solution, the active reagent has Cu(I)
   [JEE Main (June) 2022 (I)]

6. Acidified potassium permanganate solution oxidises oxalic acid. The spin-only magnetic moment of the manganese product formed from the above reaction is ______ B.M. (Nearest integer). **[JEE Main (June) 2022 (I)]**

7. The sum of oxidation state of the metals in $Fe(CO)_5$, VO^{2+} and WO_3 is ______: **[JEE Main (April) 2023 (II)]**

8. The total change in the oxidation state of manganese involved in the reaction of $KMnO_4$ and potassium iodide in the acidic medium is ______. **[JEE Main (April) 2023 (I)]**

9. In alkaline medium, the reduction of permanganate anion involves a gain of ______ electrons. **[JEE Main (April) 2023 (II)]**

10. Number of metal ions characterized by flame test among the following is ______.
    Sr^{2+}, Ba^{2+}, Ca^{2+}, Cu^{2+}, Zn^{2+}, Co^{2+}, Fe^{2+} **[JEE Main (Jan.) 2024]**

11. The molar mass of the water insoluble product formed from the fusion of chromite ore ($FeCr_2O_4$) with Na_2CO_3 in presence of O_2 is ....g mol^{-1}. **[JEE Main (Jan.) 2025]**

Answers

[A] Objective Questions (Only one option is correct)

1. (a) **2.** (c) **3.** (d) **4.** (d) **5.** (b) **6.** (b) **7.** (a) **8.** (c) **9.** (b) **10.** (d)
11. (d) **12.** (c) **13.** (b) **14.** (a) **15.** (c) **16.** (c) **17.** (d) **18.** (d) **19.** (d) **20.** (d)
21. (d) **22.** (a) **23.** (a) **24.** (b) **25.** (a) **26.** (c) **27.** (b) **28.** (c) **29.** (d) **30.** (b)
31. (a) **32.** (a) **33.** (b) **34.** (d) **35.** (b) **36.** (b) **37.** (a) **38.** (b) **39.** (b) **40.** (a)
41. (c) **42.** (c) **43.** (c) **44.** (c)

[B] Objective Questions (One or more than one option(s) is/are correct)

1. (a, d) **2.** (a, b, c) **3.** (a, b) **4.** (a, b, c) **5.** (a, c, d)

[C] Passage Based Problems

Passage-1

1. (b) **2.** (a) **3.** (c)

[D] Integer Type Problems

1. (4) **2.** (4) **3.** (19) **4.** (+6) **5.** (3)
6. (6) **7.** (10) **8.** (5) **9.** (3) **10.** (4)
11. (160)

Hints and Solutions

A. Objective Questions

[Only one option is correct]

1. (a) Number of unpaired e^- in Fe^{2+} is less than Mn^{2+}.

2. (c) $2MnO_4^- + 5C_2O_4^{2-} + 16H^+ \longrightarrow 2Mn^{2+} + 10CO_2 + 8H_2O$

3. (d) Factual.

4. (d) $Fe \xrightarrow[\text{Heat}]{O_2} Fe_3O_4$

This reaction is corresponding to the combustion of Fe.

$$Fe_3O_4 \xrightarrow[600°C]{CO} FeO \xrightarrow[700°C]{CO} Fe$$

These reactions correspond to the production of Fe by reduction of Fe_3O_4 in blast furnace.

5. (b) CrO_2 is metallic and ferromagnetic.

6. (b) $La^{3+}(4f^0)$, $Gd^{3+}(4f^7)$, $Lu^{3+}(4f^{14})$ — Colourless

$Sm^{3+}(4f^5) \Rightarrow$ Yellow coloured

7. (a)

8. (c)

Manganate ion

$\overset{+6}{MnO_4^{2-}}$

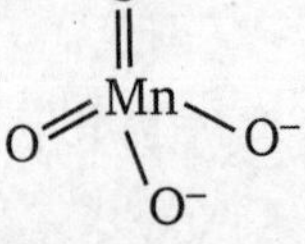

Electronic configuration $[Ar]3d^1$

Paramagnetic, green in colour, sp^3-hybridisation $d\pi$-$p\pi$ bond is present.

Permanganate ion

$\overset{+7}{MnO_4^-}$

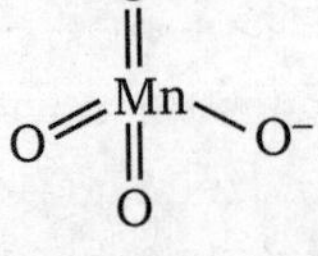

Electronic configuration [Ar]

Diamagnetic, purple in colour, sp^3-hybridisation $d\pi$-$p\pi$ bond is present.

9. (b) In $SF_4 \rightarrow$ hybridisation $= sp^3d$

Bond pair = 4

Lone pair = 1

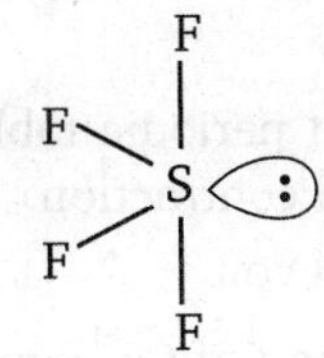

Shape: trigonal bipyramidal.

10. (d) Assertion is true but reason is false. Maximum catalytic activity is due to weak adsorption of reactant molecules over its surface.

11. (d) Carbon tetrachloride being molecular solid have weak van der Waals' forces. So that it melts at low temperature and it is bad conductor of electricity in both liquid and solid state.

12. (c) Formation of +6 oxidation state is more difficult in Cr than W reason is in chromium (Cr) electrons are tightly bounded with nucleus due to strong attractive force.

Statement (b) is incorrect. Since, permanganate ion (MnO_4^-) oxidised Cl^- to Cl_2, so, HCl is never used with $KMnO_4$ in volumetric analysis.

13. (b) $_{63}Eu$ configuration is : $[Xe]4f^7 6s^2$

Eu^{2+} configuration will be : $[Xe]4f^7$

$_{58}Ce$ configuration is : $[Xe]4f^1 5d^1 6s^2$

Ce^{3+} configuration will be : $[Xe]4f^1$

14. (a) $_{64}Gd^{3+}$ configuration is $[Xe]4f^7$

Magnetic moment $= \sqrt{n(n+2)}$

$n = 7$ (unpaired electrons)

$\mu = \sqrt{7(7+2)} = 7.9$

15. (c)

Element	Electronic configuration
$_{63}Eu$	$[Xe]4f^7 6s^2$
$_{66}Dy$	$[Xe]4f^{10} 6s^2$
$_{65}Tb$	$[Xe]4f^9 6s^2$
$_{58}Ce$	$[Xe]4f^1 5d^1 6s^2$

Ce^{4+} forms noble configuration after removing 4 electrons. Dy^{4+} and Tb^{4+} are stable since they form stable half filled orbital.

16. (c)
- ❖ For lanthanides +3 oxidation state most stable oxidation state. In CeO_2, Ce exist in +4 oxidation state. And have tendency to go to +3 oxidation state that is the reason it will act as oxidising agent for aldehyde and ketone.
- ❖ Same goes for Eu^{2+} oxidation state where it will loose one more electron and convert to Eu^{3+}. So, acts as reducing agent.

17. (d) Yb does not form MO_2 types oxide.

18. (d) Size of $_{97}Bk^{3+}$ is less than that of $_{93}Np^{3+}$.

due to actinoid contraction.

As we move from left to right in periodic table size decreases. As both are present in actinoid series they show actinoid contraction.

19. (d) $Ce^{4+} + e^- \longrightarrow Ce^{3+}$; $E° = 1.74$ volt

Since $E°$ value is positive, reduction of Ce^{4+} into Ce^{3+} is spontaneous, so Ce^{3+} will be more stable oxidation state.

20. (d)

Element	Electrode potential
Zn	−0.76 V
Fe	−0.44 V
Co	−0.28 V
Cu	+0.34 V

21. (d) Red colour of ruby is due to presence of Cr^{3+} ion.

22. (a) Statement (I) is correct

$$\underset{\text{Potassium permanganate}}{2KMnO_4} \xrightarrow{573K} \underset{\text{Potassium manganate}}{K_2MnO_4} + MnO_2 + O_2$$

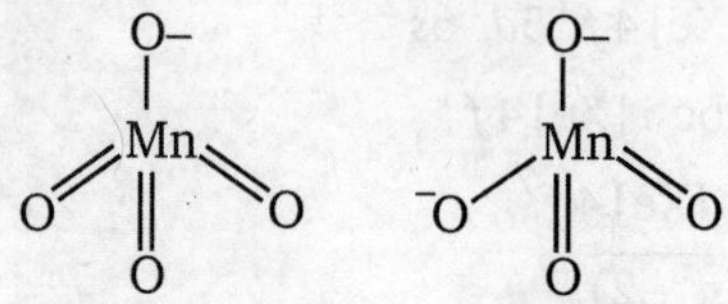

(MnO_4^-) (Tetrahedral) (MnO_4^{2-}) (Tetrahedral)

Mn electronic configuration → $[Ar]3d^5 4s^2$

In MnO_4^-, Mn is in +7 oxidation state.

In MnO_4^{2-}, Mn is in +6 oxidation state.

$$Mn^{7+} \longrightarrow [Ar]$$
$$[Mn^{6+}] \longrightarrow [Ar]3d^1$$

Since, MnO_4^- have no unpaired electron it is diamagnetic while MnO_4^{2-} have unpaired electron, it is paramangnatic.

So, statement (II) is false.

23. (a) FeI_3 is unstable due to large size of I^- and reducing nature of $I^{\ominus}$.

$$\underset{\text{Unstable}}{2FeI_3} \longrightarrow \underset{\text{Stable}}{2FeI_2 + I_2}$$

24. (b) Chlorophyll is a complex of magnesium (Mg).

Vitamin-B_{12} is a complex of cobalt (Co).

Cis-platin (Complex of platinum) is an anticancer drug.

Grubbs catalyst is a series of catalysts containing Ruthenium.

25. (a) Sc^{3+} is with d^0 and Zn^{2+} is with d^{10} configuration.

26. (c) In manganese heptoxide, there are 6, Mn=O bonds.

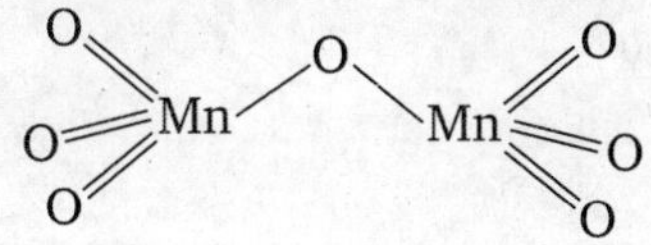

27. (b) Copper get reduced and hydrogen get oxidised $CuO + H_2 \longrightarrow Cu + H_2O$ (under hot conditions)

28. (c) $Eu \longrightarrow [Xe]4f^7 6s^2$ $\quad$ $Tm \longrightarrow [Xe]4f^{13}6s^2$

$Eu^{2+} \longrightarrow [Xe]4f^7$ $\quad$ $Tm^{2+} \longrightarrow [Xe]4f^{13}$

$Sm \longrightarrow [Xe]4f^6 6s^2$ $\quad$ $Tb \longrightarrow [Xe]4f^9 6s^2$

$Sm^{2+} \longrightarrow [Xe]4f^6$ $\quad$ $Tb^{2+} \longrightarrow [Xe]4f^9$

29. (d) In CrO_2Cl_2 oxidation state of Cr is +6

$Cr(VI) = [Ar]^{18}3d^0$, $\quad Mn(VII) = [Ar]^{18}3d^0$, $\quad Fe(III) = [Ar]^{18}3d^5$,

$Ti(III) = [Ar]^{18}3d^1$, $\quad V(IV) = [Ar]^{18}3d^1$

Hence Cr(VI) and Mn(VII) have same d^0 configuration.

30. (b) Sc does not show +4 oxidation state.

31. (a) Prolonged heating will cause oxidation of Fe^{2+} to Fe^{3+}

32. (a) With increase in oxygen acidic nature of oxide of an element increase and basic nature decreases.

33. (b) $\left.\begin{matrix} Eu^{2+} : [Xe]4f^7 \\ Yb^{2+} : [Xe]4f^{14} \end{matrix}\right\}$ High IE due to half filled & fully filled configurations

34. (d) Ce^{4+} has empty 4*f* subshell so it gains electrons thus, strongly oxidising.

35. (b) Statement (1) is true, Ce^{4+} has noble gas electronic configuration.

Statement (2) is also true due to high reduction potential for Ce^{4+}/Ce^{3+} (+1.74V), and stability of Ce^{3+}, Ce^{4+} acts as strong oxidizing agent.

36. (b) $[Cr] = [Ar]\ 4s^1 3d^5$

$[Cd] = [Kr]\ 5s^2 4d^{10}$

$[Cu] = [Ar]\ 4s^1 3d^{10}$

$[Ag] = [Kr]\ 5s^1 4d^{10}$

$[Zn] = [Ar]\ 4s^2 3d^{10}$

37. (a) (A) Zn, Cd, Hg exhibit lowest enthalpy of atomization in respective transition series.

(C) Compounds of Zn, Cd and Hg are diamagnetic in nature.

38. (b) Ce : $[Xe]\ 4f^1 5d^1 6s^2$; Ce^{4+} diamagnetic

La : $[Xe]\ 4f^0 5d^1 6s^2$; La^{3+} diamagnetic

39. (b) K_2MnO_4

$2 + x - 8 = 0$

$\Rightarrow \quad x = +6$

O.S. of Mn = +6

IUPAC Name = Potassium tetraoxidomanganate(VI)

40. (a) $_{63}Eu^{2+}$ – [Xe] $4f^7 6s^0$

$_{64}Gd^{3+}$ – [Xe] $4f^7 5d^0 6s^0$

$_{63}Eu^{3+}$ – [Xe] $4f^6 6s^0$

$_{65}Tb^{3+}$ – [Xe] $4f^8 6s^0$

$_{62}Sm^{2+}$ – [Xe] $4f^6 6s^0$

Eu^{2+} and Gd^{3+}

41. (c) $K_2Cr_2O_7 \xrightarrow[-H_2O]{KOH} \underset{[A]}{K_2CrO_4} \xrightarrow[-H_2O]{H_2SO_4} \underset{[B]}{K_2Cr_2O_7} + K_2SO_4$

42. (c) Tb^{4+} is strongest oxidising agent.

43. (c) V_2O_5 + alkali $\rightarrow VO_4^{3-}$

In VO_4^{3-} ion vanadium is in +5 oxidation state.

44. (c) M.P. $\Rightarrow$ Mn < Fe, Tc < Ru, Os < Re

B. Objective Questions

[One or more than one option(s) is/are correct]

1. (a, d) In acidic medium

$$MnO_4^- + 8H^+ + 5e^- \longrightarrow Mn^{2+} + 4H_2O$$

In neutral medium

$$MnO_4^- + 2H_2O + 3e^- \longrightarrow MnO_2 + 4OH^-$$

Hence, number of electrons loose in acidic and neutral medium are 5 and 3 electrons respectively.

2. (a, b, c)

(a) Cr^{2+} is a reducing agent because Cr^{3+} is more stable.

(b) Mn^{3+} is an oxidizing agent because Mn^{2+} is more stable.

(c) Cr^{2+} and Mn^{3+} exhibit d^4 electronic configuration.

3. (a, b)

(a) H_2O_2 in alkaline medium reduces Fe^{3+} to Fe^{2+}.

$$2Fe^{3+} + H_2O_2 + 2OH^- \longrightarrow 2Fe^{2+} + 2H_2O + O_2$$

(b) $Na_2O_2 + 2H_2O \longrightarrow H_2O_2 + 2NaOH$

(c) H_2O_2 in acidic medium oxidises Fe^{2+} to Fe^{3+}.

$$2Fe^{2+} + H_2O_2 + 2H^+ \longrightarrow 2Fe^{3+} + 2H_2O$$

(d) Na_2O_2 in presence of H_2SO_4,

$$Na_2O_2 + H_2SO_4 \longrightarrow Na_2SO_4 + \underset{\text{(neutral medium)}}{H_2O_2}$$

In alkaline medium, reducing action of H_2O_2 is more effective.

4. (a, b, c)

$$2MnO_2 + 4KOH + O_2 \xrightarrow{\text{Fusion}} \underset{(W)}{2K_2MnO_4} + 2H_2O$$

On electrolysis of K_2MnO_4, MnO_4^{2-} gets oxidised to MnO_4^-.

$$\underset{(W)}{2K_2MnO_4} + 2H_2O \xrightarrow{\text{Electrolysis}} \underset{(X)}{2KMnO_4} + 2KOH + H_2$$

K_2MnO_4 in neutral or acidic medium undergoes disproportionation reaction.

$$\underset{(Y)}{3\overset{+6}{Mn}O_4^{2-}} + 4H^+ \longrightarrow \underset{(Z)}{2\overset{+7}{Mn}O_4^-} + \overset{+4}{Mn}O_2 + 2H_2O$$

K_2MnO_4 is green in colour and $KMnO_4$ is purple in colour. Both are tetrahedral in shape involving $p\pi - d\pi$ bonding.

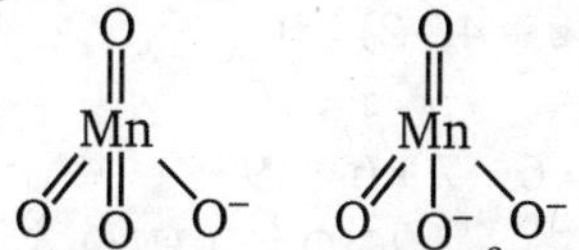

(MnO_4^-) Z (MnO_4^{2-}) Y

$\overset{+6}{Y(MnO_4^{2-})}$ is paramagnetic while $\overset{+7}{Z(MnO_4^-)}$ is diamagnetic in nature.

5. (a, c, d)

$$Zn + \underset{\text{(Hot \& conc.)}}{H_2SO_4} \longrightarrow \underset{(G)}{ZnSO_4} + \underset{(R)}{SO_2} + \underset{(X)}{H_2O}$$

$$Zn + \text{conc. NaOH} \longrightarrow \underset{(T)}{Na_2ZnO_2} + \underset{(Q)}{H_2}$$

$$\underset{(G)}{ZnSO_4} + H_2S + NH_4OH(aq.) \longrightarrow \underset{(Z)}{ZnS\downarrow} + \underset{(Y)}{(NH_4)_2SO_4} + \underset{(X)}{H_2O}$$

C. Passage Based Problems

Passage-1

1. (b) $\underset{M}{Cu} + \underset{N}{2AgNO_3} \rightarrow \underset{\text{Blue}}{Cu(NO_3)_2} + 2Ag$

While Cu partially oxidizes to $Cu(NO_3)_2$ and remaining $AgNO_3$ reacts with NaCl.

2. (a)

3. (c)

$$\underset{(N)}{AgNO_3} + NaCl \longrightarrow \underset{(O)}{AgCl\downarrow} + NaNO_3$$

$$AgCl + 2NH_3 \longrightarrow [Ag(NH_3)_2]^+Cl^-$$

$$Cu(NO_3)_2 + 4NH_4OH \longrightarrow [Cu(NH_3)_4]^{2+}$$

D. Integer Type Problems

1. (4) $CuSO_4 \cdot 5H_2O \longrightarrow [Cu(H_2O)_4]SO_4 \cdot H_2O$

So, water molecules directly attached to Cu are 4.

2. (4) $K_2CrO_4 + H_2O_2 \xrightarrow[\text{(In acidic medium)}]{\text{Amyl alcohol}} \underset{\substack{(X)\\ \text{(Blue liquid)}}}{CrO_5}$

Here the structure of CrO_5 is :

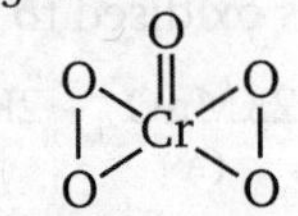

Here, single bonded O-atoms with Cr = 4

3. (19) In $K_2Cr_2O_7$, $\quad 2(+1) \times 2x + 7(-2) = 0$

$x = +6$

In $KMnO_4$, $\quad 1 + y + 4(-2) = 0$

$y = +7$

In K_2FeO_4, $\quad 2(+1) + z + 4(-2) = 0$

$z = +6$

$x + y + z = 6 + 7 + 6 = 19$

4. (+6) Reaction is $Cr_2O_7^{2-} + 2OH^- \longrightarrow 2CrO_4^{2-} + H_2O$

Let oxidation number of Cr in CrO_4^{2-} is x.

$x + 4(-2) = -2 \Rightarrow x = 8 - 2 = +6$

5. (3) A, B, C are correct and D is incorrect because Fehling solution has Cu(II).

6. (6) $2KMnO_4 + 3H_2C_2O_4 + 3H_2SO_4 \longrightarrow K_2SO_4 + 2MnSO_4 + 10CO_2 + 8H_2O$.

Mn^{2+} has 5 unpaired electrons, therefore, the magnetic moment = $\sqrt{5(5+2)} = \sqrt{35}$ BM.

7. (10) $\overset{(0)}{Fe(CO)_5} \quad \overset{(+4)}{VO^{2+}} \quad \overset{(+6)}{WO_3}$

So, sum of oxidation state = 0 + 4 + 6 = 10

8. (5) $\quad \overset{+7}{KMnO_4} \longrightarrow \overset{+2}{Mn^{2+}}$ (In acidic medium)

Change in oxidation state of Mn = 5

9. (3) In faintly alkaline medium,

$\overset{+7}{MnO_4^-} + 3e^- + 2H_2O \longrightarrow \overset{+4}{MnO_2} + 4OH^-$

No. of electrons gained = 3

10. (4) All the following metal ions will respond to flame test.

$Sr^{2+}, Ba^{2+}, Ca^{2+}, Cu^{2+}$

11. (160)

$4FeCr_2O_4 + 8Na_2CO_3 + 7O_2 \rightarrow 8Na_2CrO_4 + 2Fe_2O_3 + 8CO_2$

Fe_2O_3 is water insoluble, so its molar mass

$\Rightarrow \quad [2 \times 56 + 3 \times 16] \Rightarrow 160$ g/mol

□□□

CHAPTER 8

Coordination Compounds

A. Objective Questions

[Only one option is correct]

1. The correct structure of ethylenediaminetetraacetic acid (EDTA) is : **(IIT 2010)**

(a) 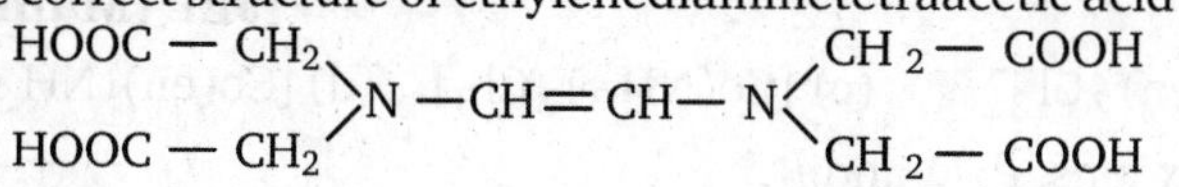

(b) 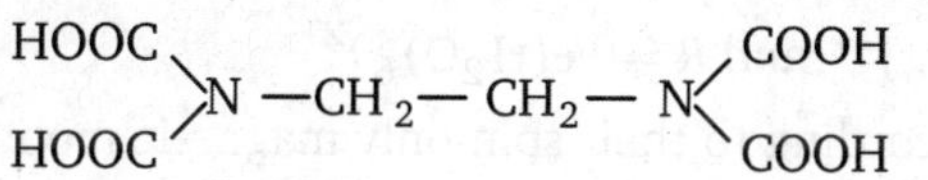

(c) 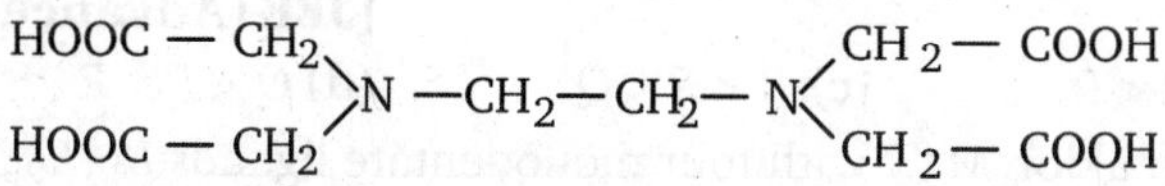

(d) 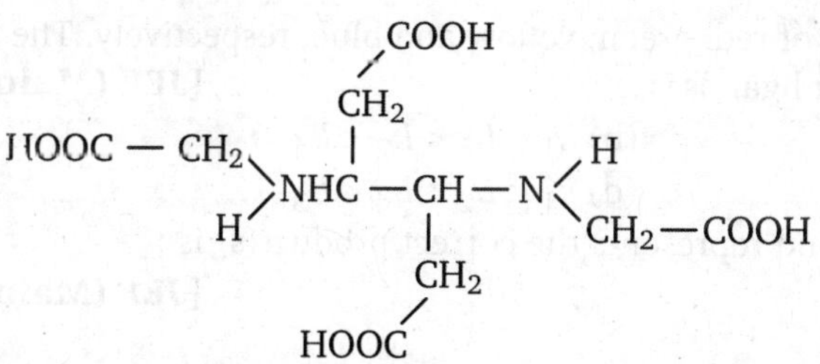

2. The ionisation isomer of $[Cr(H_2O)_4Cl(NO_2)]Cl$ is : **(IIT 2010)**

(a) $[Cr(H_2O)_4(O_2N)]Cl_2$ (b) $[Cr(H_2O)_4(Cl_2)](NO_2)$

(c) $[Cr(H_2O)_4Cl(ONO)]Cl$ (d) $[Cr(H_2O)_4Cl_2(NO_2)]H_2O$

3. The complex showing a spin-only magnetic moment of 2.83 B.M. is : **(IIT 2010)**

(a) $Ni(CO)_4$ (b) $[NiCl_4]^{2-}$ (c) $Ni(PPh_3)_4$ (d) $[Ni(CN)_4]^{2-}$

4. Geometrical shapes of the complexes formed by the reaction of Ni^{2+} with Cl^-, CN^- and H_2O, respectively, are : **(IIT 2011)**

(a) octahedral, tetrahedral and square planar

(b) tetrahedral, square planar and octahedral

(c) square planar, tetrahedral and octahedral
(d) octahedral, square planar and octahedral

5. Among the following complexes **(K – P)**
$K_3[Fe(CN)_6]$ **(K)**, $[Co(NH_3)_6]Cl_3$ **(L)**, $Na_3[Co(oxalate)_3]$ **(M)**, $[Ni(H_2O)_6]Cl_2$ **(N)**, $K_2[Pt(CN)_4]$ **(O)** and $[Zn(H_2O)_6](NO_3)_2$ **(P)** **(IIT 2011)**
(a) **K, L, M, N** (b) **K, M, O, P** (c) **L, M, O, P** (d) **L, M, N, O**

6. $NiCl_2\{P(C_2H_5)_2(C_6H_5)\}_2$ exhibits temperature dependent magnetic behaviour (paramagnetic/diamagnetic). The coordination geometries of Ni^{2+} in the paramagnetic and diamagnetic states are respectively : **(IIT 2012)**
(a) tetrahedral and tetrahedral (b) square planar and square planar
(c) tetrahedral and square planar (d) square planar and tetrahedral

7. The colour of light absorbed by an aqueous solution of $CuSO_4$ is : **(IIT 2012)**
(a) orange-red (b) blue-green (c) yellow (d) violet

8. As per IUPAC nomenclature, the name of the complex $[Co(H_2O)_4(NH_3)_2]Cl_3$ is : **(IIT 2012)**
(a) tetraaquadiaminecobalt(III) chloride (b) tetraaquadiamminecobalt(III) chloride
(c) diaminetetraaquacobalt(III) chloride (d) diamminetetraaquacobalt(III) chloride

9. Which of the following complex species is not expected to exhibit optical isomerism? **[JEE (Mains) 2013]**
(a) $[Co(en)_3]^{3+}$ (b) $[Co(en)_2Cl_2]^+$ (c) $[Co(NH_3)_3Cl_3]$ (d) $[Co(en)(NH_3)_2Cl_2]^+$

10. Consider the following complex ions, P, Q and R.
$$P = [FeF_6]^{3-}, Q = [V(H_2O)_6]^{2+} \text{ and } R = [Fe(H_2O)_6]^{2+}$$
The correct order of the complex ions, according to their spin-only magnetic moment values (in B.M.) is : **[JEE (Advanced) 2013]**
(a) $R < Q < P$ (b) $Q < R < P$ (c) $R < P < Q$ (d) $Q < P < R$

11. The octahedral complex of a metal ion M^{3+} with four monodentate ligands L_1, L_2, L_3 and L_4 absorb wavelengths in the region of red, green, yellow and blue, respectively. The increasing order of ligand strength of the four ligands is : **[JEE (Mains) 2014]**
(a) $L_4 < L_3 < L_2 < L_1$ (b) $L_1 < L_3 < L_2 < L_4$
(c) $L_3 < L_2 < L_4 < L_1$ (d) $L_1 < L_2 < L_4 < L_3$

12. The equation which is balanced and represents the correct product(s) is : **[JEE (Mains) 2014]**
(a) $Li_2O + 2KCl \longrightarrow 2LiCl + K_2O$
(b) $[CoCl(NH_3)_5]^+ + 5H^+ \longrightarrow Co^{2+} + 5NH_4^+ + Cl^-$
(c) $[Mg(H_2O)_6]^{2+} + (EDTA)^{4-} \xrightarrow{\text{Excess NaOH}} [Mg(EDTA)]^{2+} + 6H_2O$
(d) $CuSO_4 + 4KCN \longrightarrow K_2[Cu(CN)_4] + K_2SO_4$

13. Which of the following compounds is not yellow coloured ? **[JEE (Mains) 2015]**
(a) $(NH_4)_3[As(Mo_3O_{10})_4]$ (b) $BaCrO_4$
(c) $Zn_2[Fe(CN)_6]$ (d) $K_3[Co(NO_2)_6]$

14. The number of geometric isomers that can exist for square planar $[Pt(Cl)(py)(NH_3)(NH_2OH)]^+$ is (py = pyridine) **[JEE (Mains) 2015]**
(a) 4 (b) 6 (c) 2 (d) 3

15. The colour of $KMnO_4$ is due to : **[JEE (Mains) 2015]**

(a) $L \rightarrow M$ charge transfer transition (b) $\sigma \rightarrow \sigma^*$ transition

(c) $M \rightarrow L$ charge transfer transition (d) $d-d$ transition

16. The pair having the same magnetic moment is :
[At. No.: Cr =24, Mn = 25, Fe = 26, Co = 27] **[JEE (Mains) 2016]**

(a) $[Cr(H_2O)_6]^{2+}$ and $[CoCl_4]^{2-}$

(b) $[Cr(H_2O)_6]^{2+}$ and $[Fe(H_2O)_6]^{2+}$

(c) $[Mn(H_2O)_6]^{2+}$ and $[Cr(H_2O)_6]^{2+}$

(d) $[CoCl_4]^{2-}$ and $[Fe(H_2O)_6]^{2+}$

17. Which one of the following complexes shows optical isomerism ? **[JEE (Mains) 2016]**

(a) $[Co(NH_3)_3Cl_3]$

(b) *cis* $[Co(en)_2Cl_2]Cl$

(c) *trans* $[Co(en)_2Cl_2]Cl$

(d) $[Co(NH_3)_4Cl_2]Cl$ (en = ethylenediamine)

18. Among $[Ni(CO)_4]$, $[NiCl_4]^{2-}$, $[Co(NH_3)_4Cl_2]Cl$, $Na_3[CoF_6]$, Na_2O_2 and CsO_2, the total number of paramagnetic compounds is: **[JEE (Advanced) 2016]**

(a) 2 (b) 3 (c) 4 (d) 5

19. The geometries of the ammonia complexes of Ni^{2+}, Pt^{2+} and Zn^{2+}, respectively, are:
[JEE (Advanced) 2016]

(a) Octahedral, square planar and tetrahedral

(b) square planar, octahedral and tetrahedral

(c) tetrahedral, square planar and octahedral

(d) octahedral, tetrahedral and square planar.

20. On treatment of 100 mL of 0.1 *M* solution of $CoCl_3 \cdot 6H_2O$ with excess $AgNO_3$, 1.2×10^{22} ions are precipitated. The complex is: **[JEE (Mains) 2017]**

(a) $[Co(H_2O)_6]Cl_3$ (b) $[Co(H_2O)_5Cl]Cl_2 \cdot H_2O$

(c) $[Co(H_2O)_4Cl_2]Cl \cdot 2H_2O$ (d) $[Co(H_2O)_3Cl_3] \cdot 3H_2O$

21. Consider the following reaction and statements:

$$[Co(NH_3)_4Br_2]^+ + Br^- \longrightarrow [Co(NH_3)_3Br_2] + NH_3$$

(I) Two isomers are produced if the reactant complex ion is a *cis* - isomer

(II) Two isomers are produced if the reactant complex ion is a *trans*- isomer.

(III) Only one isomer is produced if the reactant complex ion is a *trans*-isomer.

(IV) Only one isomer is produced if the reactant complex ion is a *cis*- isomer.

The correct statements are: **[JEE (Mains) 2018]**

(a) (III) and (IV) (b) (II) and (IV)

(c) (I) and (II) (d) (I) and (III)

22. The oxidation states of Cr in $[Cr(H_2O)_6]Cl_3$, $[Cr(C_6H_6)_2]$ and $K_2[Cr(CN)_2(O)_2(O)_2(NH_3)]$ respectively are: **[JEE (Mains) 2018]**

(a) +3, 0 and +6 (b) +3, 0 and +4

(c) +3, +4 and +6 (d) +3, +2 and +4

23. The following ligand is : **[JEE (Mains) 2019]**

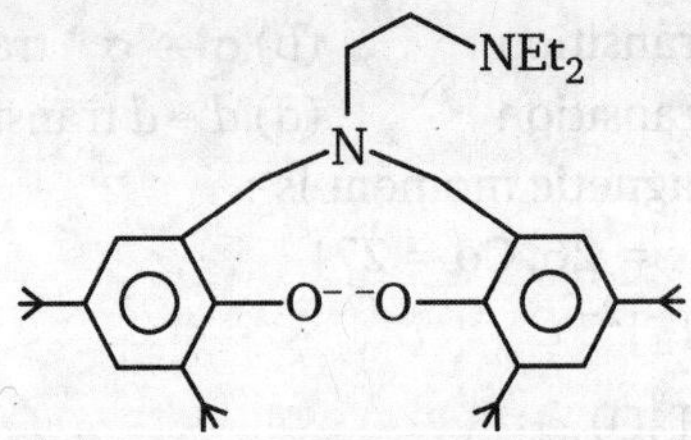

(a) hexadentate (b) tetradentate (c) bidentate (d) tridentate

24. The correct order of the spin-only magnetic moment of metal ions in the following low-spin complexes, $[V(CN)_6]^{4-}$, $[Fe(CN)_6]^{4-}$, $[Ru(NH_3)_6]^{3+}$ and $[Cr(NH_3)_6]^{2+}$ is : **[JEE (Mains) 2019]**

(a) $Cr^{2+} > Ru^{3+} > Fe^{2+} > V^{2+}$ (b) $V^{2+} > Ru^{3+} > Cr^{2+} > Fe^{2+}$
(c) $Cr^{2+} > V^{2+} > Ru^{3+} > Fe^{2+}$ (d) $V^{2+} > Cr^{2+} > Ru^{3+} > Fe^{2+}$

25. The compound that inhibits the growth of tumors is : **[JEE (Mains) 2019]**

(a) *cis*-$[Pd (Cl)_2(NH_3)_2]$ (b) *trans*-$[Pt (Cl)_2(NH_3)_2]$
(c) *trans*-$[Pd (Cl)_2(NH_3)_2]$ (d) *cis*-$[Pt (Cl)_2(NH_3)_2]$

26. The calculated spin-only magnetic moments (BM) of the anionic and cationic species of $[Fe (H_2O)_6]_2$ and $[Fe (CN)_6]$, respectively, are : **[JEE (Mains) 2019]**

(a) 0 and 5.92 (b) 4.9 and 0 (c) 2.84 and 5.92 (d) 0 and 4.9

27. The ion that has sp^3d^2 hybridization for the central atom, is : **[JEE (Mains) 2019]**

(a) $[IF_6]^-$ (b) $[ICl_4]^-$ (c) $[BrF_2]^-$ (d) $[ICl_2]^-$

28. Complex X of composition $Cr(H_2O)_6Cl_n$ has a spin only magnetic moment of 3.83 BM. It reacts with $AgNO_3$ and shows geometrical isomerism. The IUPAC nomenclature of X is: **[JEE Main (Jan.) 2020]**

(a) Dichloridotetraaqua chromium (IV) chloride dihydrate
(b) Tetraaquadichlorido chromium (III) chloride dihydrate
(c) Tetraaquadichlorido chromium (IV) chloride dihydrate
(d) Hexaaqua chromium (III) chloride

29. Complex A has a composition of $H_{12}O_6Cl_3Cr$. If the complex on treatment with conc. H_2SO_4 loses 13.5% of its original mass, the correct molecular formula of A is:
[Given: atomic mass of Cr = 52 amu and Cl = 35 amu] **[JEE Main (Sept.) 2020]**

(a) $[Cr(H_2O)_4Cl_2]Cl \cdot 2H_2O$ (b) $[Cr(H_2O)_6]Cl_3$
(c) $[Cr(H_2O)_5Cl]Cl_2 \cdot H_2O$ (d) $[Cr(H_2O)_3Cl_3] \cdot 3H_2O$

30. The IUPAC name of the complex is $[Pt(NH_3)_2Cl(NH_2CH_3)]Cl$ is: **[JEE Main (Jan.) 2020]**

(a) Diamminechlorido(aminomethane)platinum(II) chloride
(b) Diamminechlorido(methanamine)platinum(II) chloride
(c) Diammine(methanamine)chlorido platinum(II) chloride
(d) Bisammine(methanamine)chloride platinum(II) chloride

31. The complex that can show fac- and mer-isomers is: **[JEE Main (Jan.) 2020]**

(a) $[CoCl_2(en)_2]$ (b) $[Co(NH_3)_3(NO_2)_3]$

(c) $[Pt(NH_3)_2Cl_2]$ (d) $[Co(NH_3)_4Cl_2]^+$

32. The one that is not expected to show isomerism is: **[JEE Main (Sept.) 2020]**

(a) $[Ni(en)_3]^{2+}$ (b) $[Ni(NH_3)_4(H_2O)_2]^{2+}$

(c) $[Pt(NH_3)_2Cl_2]$ (d) $[Ni(NH_3)_2Cl_2]$

33. The complex that can shows optical activity is: **[JEE Main (Sept.) 2020]**

(a) *cis* - $[CrCl_2(ox)_2]^{3-}$ (ox = oxalate) (b) *trans* - $[Cr(Cl_2)(ox)_2]^{3-}$

(c) *cis* - $[Fe(NH_3)_2(CN)_4]^-$ (d) *trans* - $[Fe(NH_3)_2(CN)_4]^-$

34. Consider the complex ions, *trans* - $[Co(en)_2Cl_2]^+$ (A) and *cis* - $[Co(en)_2Cl_2]^+$ (B). The correct statement regarding them is: **[JEE Main (Sept.) 2020]**

(a) both (A) and (B) cannot be optically active.

(b) (A) cannot be optically active, but (B) can be optically active.

(c) both (A) and (B) can be optically active.

(d) (A) can be optically active, but (B) cannot be optically active.

35. The molecule in which hybrid MOs involve only one *d*-orbital of the central atom is: **[JEE Main 2020]**

(a) $[Ni(CN)_4]^{2-}$ (b) $[CrF_6]^{3-}$ (c) BrF_5 (d) XeF_4

36. Among the statements (A-D), the incorrect ones are: **[JEE Main (Jan.) 2020]**

(A) Octahedral Co(III) complexes with strong field ligands have very high magnetic moments

(B) When $\Delta_o < P$, the *d*-electron configuration of Co(III) in an octahedral complex is $t_{eg}^4 e_g^2$

(C) Wavelength of light absorbed by $[Co(en)_3]^{3+}$ is lower than that of $[CoF_6]^{3-}$

(D) If the Δ_o for an octahedral complex of Co(III) is 18,000 cm^{-1}, the Δ_t for its tetrahedral complex with the same ligand will be 16,000 cm^{-1}

(a) (B) and (C) only (b) (A) and (B) only

(c) (C) and (D) only (d) (A) and (D) only

37. The correct order of the spin-only magnetic moments of the following complexes is : **[JEE Main (Jan.) 2020]**

(I) $[Cr(H_2O)_6]Br_2$ (II) $Na_4[Fe(CN)_6]$

(III) $Na_3[Fe(C_2O_4)_3]$ $(\Delta_0 > P)$ (IV) $(Et_4N)_2[CoCl_4]$

(a) (I) > (IV) > (III) > (II) (b) (II) > (I) > (IV) > (III)

(c) (III) > (I) > (IV) > (II) (d) (III) > (I) > (II) > (IV)

38. The one that can exhibit highest paramagnetic behaviour among the following is:
gly = glycinato; bpy = 2, 2'-bipyridine **[JEE Main (Sept.) 2020]**

(a) $[Fe(en)(bpy)(NH_3)_2]^{2+}$ (b) $[Co(ox)_2(OH)_2]^-$ $(\Delta_0 > P)$

(c) $[Pd(gly)_2]$ (d) $[Ti(NH_3)_6]^{3+}$

39. The species that has a spin only magnetic moment of 5.9 BM, is: **[JEE Main (Sept.) 2020]**
(T_d = tetrahedral)

(a) $[Ni(CN)_4]^{2-}$ – (square planar) (b) $[NiCl_4]^2$ – (T_d)

(c) $Ni(CO)_4$ (T_d) (d) $[MnBr_4]^2$ – (T_d)

40. $[Pd(F)(Cl)(Br)(I)]^{2-}$ has n number of geometrical isomers. Then, the spin-only magnetic moment and crystal field stabilisation energy [CFSE] of $[Fe(CN)_6]^{n-6}$, respectively, are:
[Note: Ignore the pairing energy] **[JEE Main (Jan.) 2020]**

(a) 2.84 BM and $-1.6\,\Delta_0$ (b) 1.73 BM and $-2.0\,\Delta_0$
(c) 5.92 BM and 0 (d) 0 BM and $-2.4\,\Delta_0$

41. For octahedral Mn(II) and tetrahedral Ni(II) complexes, consider the following statements:
(I) both the complexes can be high spin.
(II) Ni(II) complex can very rarely be of low spin.
(III) with strong field ligands, Mn(II) complexes can be low spin.
(IV) aqueous solution of Mn(II) ions is yellow in colour.
The correct statements are : **[JEE Main (Sept.) 2020]**

(a) (II), (III) and (IV) only (b) (I), (II) and (III) only
(c) (I), (III) and (IV) only (d) (I) and (II) only

42. Consider that d^6 metal ion (M^{2+}) forms a complex with aqua ligands, and the spin only magnetic moment of the complex is 4.90 BM. The geometry and the crystal field stabilization energy of the complex is: **[JEE Main (Sept.) 2020]**

(a) tetrahedral and $-1.6\Delta_t + 1P$ (b) tetrahedral and $-0.6\Delta_t$
(c) octahedral and $-2.4\Delta_o + 2P$ (d) octahedral and $-1.6\,\Delta_o$

43. Simplified absorption spectra of three complexes [(i) and (ii) and (iii)] of M^{n+} ion are provided below; their λ_{max} values are marked as A, B and C respectively. The correct match between the complexes and their λ_{max} values is: **[JEE Main (Sept.) 2020]**

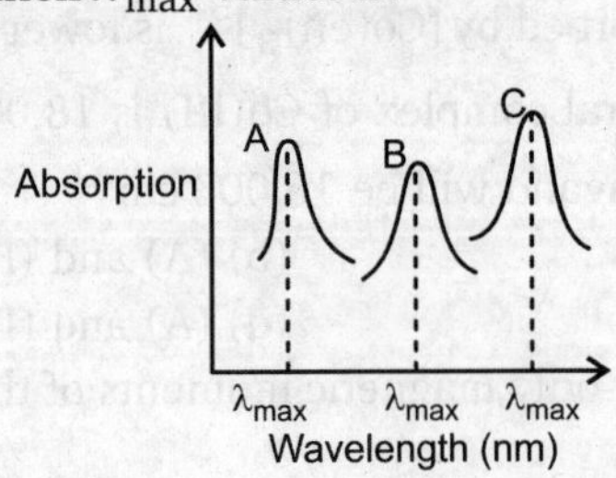

(i) $[M(NCS)_6]^{(-6+n)}$ (ii) $[MF_6]^{(-6+n)}$ (iii) $[M(NH_3)_6]^{n+}$

(a) $A \to$ i; $B \to$ ii; $C \to$ iii (b) $A \to$ ii; $B \to$ iii; $C \to$ i
(c) $A \to$ iii; $B \to$ i; $C \to$ ii (d) $A \to$ ii; $B \to$ i; $C \to$ iii

44. The d-electron configuration of $[Ru(en)_3]Cl_2$ and $[Fe(H_2O)_6]Cl_2$, respectively are: **[JEE Main (Sept.) 2020]**

(a) $t_{2g}^6 e_g^0$ and $t_{2g}^4 e_g^2$ (b) $t_{2g}^4 e_g^2$ and $t_{2g}^4 e_g^2$ (c) $t_{2g}^6 e_e^0$ and $t_{2g}^6 e_g^0$ (d) $t_{2g}^4 e_g^2$ and $t_{2g}^4 e_g^0$

45. The Crystal Field Stabilization Energy (CFSE) of $[CoF_3(H_2O)_3]$ $(\Delta_0 < P)$ is: **[JEE Main (Sept.) 2020]**

(a) $-0.8\,\Delta_0$ (b) $-0.4\,\Delta_0 + P$ (c) $-0.8\,\Delta_0 + 2P$ (d) $-0.4\,\Delta_0$

46. The values of the crystal field stabilization energies for a high spin d^6 metal ion in octahedral and tetrahedral fields, respectively, are: **[JEE Main (Sept.) 2020]**

(a) $-0.4\,\Delta_0$ and $-0.27\,\Delta_t$ (b) $-1.6\,\Delta_0$ and $-0.4\,\Delta_t$
(c) $-0.4\,\Delta_0$ and $-0.6\,\Delta_t$ (d) $-2.4\,\Delta_0$ and $-0.6\,\Delta_t$

47. For a d^4 metal ion in an octahedral field, the correct electronic configuration is: **[JEE Main (Sept.) 2020]**

(a) $t_{2g}^3 e_g^1$ when $\Delta_0 < P$
(b) $t_{2g}^3 e_g^1$ when $\Delta_0 > P$
(c) $t_{2g}^4 e_g^0$ when $\Delta_0 < P$
(d) $t_g^2 e_{2g}^2$ when $\Delta_0 < P$

48. Given below are two statements :

Statement I : The identification of Ni^{2+} is carried out by dimethyl glyoxime in the presence of NH_4OH.

Statement II : The dimethyl glyoxime is a bidentate neutral ligand.

In the light of the above statements, choose the correct answer from the options given below. **[JEE Main (Feb.) 2021]**

(a) Statement I is false but Statement II is true.
(b) Both Statement I and Statement II are false.
(c) Statement I is true but Statement II is false.
(d) Both Statement I and Statement II are true.

49. In which of the following order the given complex ions are arranged correctly with respect to their decreasing spin only magnetic moment? **[JEE Main (Feb.) 2021]**

(i) $[FeF_6]^{3-}$
(ii) $[Co(NH_3)_6]^{3+}$
(iii) $[NiCl_4]^{2-}$
(iv) $[Cu(NH_3)_4]^{2+}$

(a) (i) > (iii) > (iv) > (ii)
(b) (ii) > (iii) > (i) > (iv)
(c) (iii) > (iv) > (ii) > (i)
(d) (ii) > (i) > (iii) > (iv)

50. The hybridization and magnetic nature of $[Mn(CN)_6]^{4-}$ and $[Fe(CN)_6]^{3-}$, respectively are: **[JEE Main (Feb.) 2021]**

(a) d^2sp^3 and diamagnetic
(b) sp^3d^2 and diamagnetic
(c) d^2sp^3 and paramagnetic
(d) sp^3d^2 and paramagnetic

51. Correct formula of the compound which gives a white precipitate with $BaCl_2$ solution, but not with $AgNO_3$ solution, is : **[JEE Main (June) 2022 (I)]**

(a) $[Co(NH_3)_5Br]SO_4$
(b) $[Co(NH_3)_5SO_4]Br$
(c) $[Pt(NH_3)_4Cl_2]Br_2$
(d) $[Pt(NH_3)_4Br_2]Cl_2$

52. White precipitate of AgCl dissolves in aqueous ammonia solution due to formation of : **[JEE Main (June) 2022 (I)]**

(a) $[Ag(NH_3)_4]Cl_2$
(b) $[Ag(Cl_2)(NH_3)_2]$
(c) $[Ag(NH_3)_2]Cl$
(d) $[Ag(NH_3)Cl]Cl$

53. Octahedral complexes of copper (II) undergo structural distortion (Jahn-Teller). Which one of the given copper (II) complexes will show the maximum structural distortion?
(en-ethylenediamine; $H_2N—CH_2—CH_2—NH_2$) **[JEE Main (July) 2022 (II)]**

(a) $[Cu(H_2O)_6]SO_4$
(b) $[Cu(en)(H_2O)_4]SO_4$
(c) *cis*-$[Cu(en)_2Cl_2]$
(d) *trans*-$[Cu(en)_2Cl_2]$

54. Low oxidation state of metals in their complexes are common when ligands : **[JEE Main (July) 2022 (II)]**

(a) have good π-accepting character (b) have good σ-donor character
(c) are having good π-donating ability (d) are having poor σ-donating ability

55. The correct order of energy of absorption for the following metal complexes is :
A : $[Ni(en)_3]^{2+}$, B : $[Ni(NH_3)_6]^{2+}$, C : $[Ni(H_2O)_6]^{2+}$ **[JEE Main (July) 2022 (II)]**
(a) C < B < A (b) B < C < A (c) C < A < B (d) A < C < B

56. Which of the following are the example of double salt?
(A) $FeSO_4 \cdot (NH_4)_2SO_4 \cdot 6H_2O$ (B) $CuSO_4 \cdot 4NH_3 \cdot H_2O$
(C) $K_2SO_4 \cdot Al_2(SO_4)_3 \cdot 24H_2O$ (D) $Fe(CN)_2 \cdot 4KCN$
Choose the correct answer : **[JEE Main (Feb.) 2023 (I)]**
(a) A, B and D only (b) A and B only
(c) B and D only (d) A and C only

57. The complex cation which has two isomers is : **[JEE Main (Feb.) 2023 (II)]**
(a) $[Co(NH_3)_5Cl]^{2+}$ (b) $[Co(NH_3)_5Cl]^{+}$
(c) $[Co(NH_3)_5NO_2]^{2+}$ (d) $[Co(H_2O)_6]^{3+}$

58. Which of the following complex has a possibility to exist as meridional isomer?
[JEE Main (April) 2023 (I)]
(a) $[Co(NH_3)_3(NO_2)_3]$ (b) $[Co(en)_3]$
(c) $[Co(en)_2Cl_2]$ (d) $[Pt(NH_3)_2Cl_2]$

59. The complex with highest magnitude of crystal field splitting energy (Δ_0) is :
[JEE Main (April) 2023 (I)]
(a) $[Cr(OH_2)_6]^{3+}$ (b) $[Ti(OH_2)_6]^{3+}$ (c) $[Fe(OH_2)_6]^{3+}$ (d) $[Mn(OH_2)_6]^{3+}$

60. If Ni^{2+} is replaced by Pt^{2+} in the complex $[NiCl_2Br_2]^{2-}$, which of the following properties are expected to get changed? **[JEE Main (April) 2023 (II)]**
(A) Geometry (B) Geometrical isomerism
(C) Optical isomerism (D) Magnetic properties
(a) A, B and C (b) A, B and D
(c) A and D (d) B and C

61. The correct order of the number of unpaired electrons in the given complexes is :
[JEE Main (April) 2023 (II)]
(A) $[Fe(CN)_6]^{3-}$ (B) $[FeF_6]^{3-}$
(C) $[CoF_6]^{3-}$ (D) $[Cr(oxalate)_3]^{3-}$
(E) $[Ni(CO)_4]$
Choose the correct answer from the options given below :
(a) A < E < D < C < B (b) E < A < D < C < B
(c) E < A < B < D < C (d) A < E < C < B < D

62. The correct statements from following are :
A. The strength of anionic ligands can be explained by crystal field theory
B. Valence bond theory does not give a quantitative interpretation of kinetic stability of coordination compounds.

C. The hybridization involved in formation of $[Ni(CN)_4]^{2-}$ complex is dsp^2.

D. The number of possible isomer(s) of *cis*-$[PtCl_2(en)_2]^{2+}$ is one

Choose the correct answer from the options given below : **[JEE Main (Jan.) 2024]**

(a) A, D only (b) A, C only (c) B, D only (d) B, C only

63. A reagent which gives brilliant red precipitate with nickel ions in basic medium is : **[JEE Main (Jan.) 2024]**

(a) sodium nitroprusside (b) neutral $FeCl_3$
(c) meta-dinitrobenzene (d) dimethyl glyoxime

64. The orange colour of $K_2Cr_2O_7$ and purple colour of $KMnO_4$ is due to : **[JEE Main (Jan.) 2024]**

(a) Charge transfer transition in both.
(b) $d \rightarrow d$ transition in $KMnO_4$ and charge transfer transitions in $K_2Cr_2O_7$
(c) $d \rightarrow d$ transition in $K_2Cr_2O_7$ and charge transfer transition in $KMnO_4$.
(d) $d \rightarrow d$ transition in both.

65. The metals that are employed in the battery industries are :

A. Fe B. Mn C. Ni D. Cr
E. Cd

Choose the correct answer from the options given below : **[JEE Main (Jan.) 2024]**

(a) B, C and E only (b) A, B, C, D and E
(c) A, B, C and D only (d) B, D and E only

66. Choose the correct statements from the following :

(A) Ethane-1,2-diamine is a chelating ligands
(B) Metallic aluminium is produced by electrolysis of aluminium oxide in presence of cryolite.
(C) Cyanide ion is used as ligand for leaching of silver.
(D) Phosphine act as a ligand in Wilkinson catalyst
(E) The stability constants of Ca^{2+} and Mg^{2+} are similar with EDTA complexes.

Choose the correct answer from the options given below : **[JEE Main (Jan.) 2024]**

(a) (B), (C), (E) only (b) (C), (D), (E) only
(c) (A), (B), (C) only (d) (A), (D), (E) only

67. Aluminium chloride in acidified aqueous solution forms an ion having geometry : **[JEE Main (Jan.) 2024]**

(a) Octahedral (b) Square Planar
(c) Tetrahedral (d) Trigonal bipyramidal

68. In which of the following complexes the CFSE, Δ_0 will be equal to zero? **[JEE Main (Jan.) 2025]**

(a) $[Fe(NH_3)_6]Br_2$ (b) $[Fe(en)_3]Cl_3$ (c) $K_4[Fe(CN)_6]$ (d) $K_3[Fe(SCN)_6]$

69. From the magnetic behaviour of $[NiCl_4]^{2-}$ (paramagnetic) and $[Ni(CO)_4]$ (diamagnetic), choose the correct geometry and oxidation state. **[JEE Main (Jan.) 2025]**

(a) $[NiCl_4]^{2-}$: Ni^{II}, square planar; $[Ni(CO)_4]$:Ni(0), square planar
(b) $[NiCl_4]^{2-}$: Ni^{II}, tetrahedral; $[Ni(CO)_4]$:Ni(0), tetrahedral

(c) $[NiCl_4]^{2-}$: Ni^{II}, tetrahedral; $[Ni(CO)_4]$: Ni^{II}, square planar

(d) $[NiCl_4]^{2-}$: Ni(0), tetrahedral; $[Ni(CO)_4]$: Ni(0), square planar

70. The correct order of the following complexes in terms of their crystal field stabilization energies is : **[JEE Main (Jan.) 2025]**

(a) $[Co(NH_3)_4]^{2+} < [Co(NH_3)_6]^{2+} < [Co(en)_3]^{3+} < [Co(NH_3)_6]^{3+}$

(b) $[Co(NH_3)_4]^{2+} < [Co(NH_3)_6]^{2+} < [Co(NH_3)_6]^{3+} < [Co(en)_3]^{3+}$

(c) $[Co(NH_3)_6]^{2+} < [Co(NH_3)_6]^{3+} < [Co(NH_3)_4]^{2+} < [Co(en)_3]^{3+}$

(d) $[Co(en)_3]^{3+} < [Co(NH_3)_6]^{3+} < [Co(NH_3)_6]^{2+} < [Co(NH_3)_4]^{2+}$

71. Identify the homoleptic complex(es) that is/are low spin.

(A) $[Fe(CN)_5NO]^{2-}$ (B) $[CoF_6]^{3-}$ (C) $[Fe(CN)_6]^{4-}$ (D) $[Co(NH_3)_6]^{3+}$

(E) $[Cr(H_2O)_6]^{2+}$

Choose the correct answer from the options given below : **[JEE Main (Jan.) 2025]**

(a) (B) and (E) only (b) (A) and (C) only (c) (C) and (D) only (d) (C) only

72. The *d*-electronic configuration of an octahedral Co(II) complex having magnetic moment of 3.95 BM is : **[JEE Main (Jan.) 2025]**

(a) $t_{2g}^6 e_g^1$ (b) $t_{2g}^3 e_g^0$ (c) $t_{2g}^5 e_g^2$ (d) $e^4 t_2^3$

73. The complex that shows Facial-Meridional isomerism is : **[JEE Main (Jan.) 2025]**

(a) $[Co(NH_3)_3Cl_3]$ (b) $[Co(NH_3)_4Cl_2]^+$ (c) $[Co(en)_3]^{3+}$ (d) $[Co(en)_2Cl_2]^+$

74. The correct set of ions (aqueous solution) with same colour from the following is : **[JEE Main (Jan.) 2025]**

(a) V^{2+}, Cr^{3+}, Mn^{3+} (b) Zn^{2+}, V^{3+}, Fe^{3+} (c) Ti^{4+}, V^{4+}, Mn^{2+} (d) $Sc^{3+}, Ti^{3+}, Cr^{2+}$

75. Identify the coordination complexes in which the central metal ion has d^4 configuration.

(A) $[FeO_4]^{2-}$ (B) $[Mn(CN)_6]^{3-}$ (C) $[Fe(CN)_6]^{3-}$

(D) $Cr_2(O-\overset{\overset{O}{\|}}{C}-Me)_4(H_2O)_2$ (E) $[NiF_6]^{2-}$

Choose the correct answer from the options given below : **[JEE Main (Jan.) 2025]**

(a) (C) and (E) only (b) (B), (C) and (D) only

(c) (B) and (D) only (d) (A), (B) and (E) only

76. One mole of the octahedral complex compound $Co(NH_3)_5Cl_3$ gives 3 moles of ions on dissolution in water. One mole of the same complex reacts with excess of $AgNO_3$ solution to yield two moles of $AgCl_{(s)}$. The structure of the complex is : **[JEE Main (Jan.) 2025]**

(a) $[Co(NH_3)_5Cl]Cl_2$ (b) $[Co(NH_3)_4Cl]\cdot Cl_2 \cdot NH_3$

(c) $[Co(NH_3)_4Cl_2]Cl \cdot NH_3$ (d) $[Co(NH_3)_3Cl_3]\cdot 2NH_3$

77. Find the compound '*A*' from the following reaction sequences. **[JEE Main (Jan.) 2025]**

$$A \xrightarrow{\text{aqua-regia}} B \xrightarrow[\text{(2) AcOH}]{\text{(1) } KNO_2/NH_4OH} \text{yellow ppt.}$$

(a) ZnS (b) CoS (c) MnS (d) NiS

78. When Ethane-1,2-diamine is added progressively to an aqueous solution of Nickel (II) chloride, the sequence of colour change observed will be : **[JEE Main (Jan.) 2025]**

(a) Pale Blue → Blue → Green → Violet (b) Pale Blue → Blue → Violet → Green
(c) Green → Pale Blue → Blue → Violet (d) Violet → Blue → Pale Blue → Green

79. The conditions and consequence that favours the t_{2g}^3, e_g^1 configuration in a metal complex are : **[JEE Main (Jan.) 2025]**

(a) weak field ligand, high spin complex (b) strong field ligand, high spin complex
(c) strong field ligand, low spin complex (d) weak field ligand, low spin complex

80. The correct increasing order of stability of the complexes based on Δ_o value is :

(I) $[Mn(CN)_6]^{3-}$ (II) $[Co(CN)_6]^{4-}$ (III) $[Fe(CN)_6]^{4-}$ (IV) $[Fe(CN)_6]^{3-}$

[JEE Main (Jan.) 2025]

(a) II < III < I < IV (b) IV < III < II < I (c) I < II < IV < III (d) III < II < IV < I

81. The calculated spin-only magnetic moments of $K_3[Fe(OH)_6]$ and $K_4[Fe(OH)_6]$ respectively are : **[JEE Main (Jan.) 2025]**

(a) 4.90 and 4.90 B.M. (b) 5.92 and 4.90 B.M.
(c) 3.87 and 4.90 B.M. (d) 4.90 and 5.92 B.M.

82. Identify the homoleptic complexes with odd number of *d* electrons in the central metal.

(A) $[FeO_4]^{2-}$ (B) $[Fe(CN)_6]^{3-}$ (C) $[Fe(CN)_5NO]^{2-}$ (D) $[CoCl_4]^{2-}$
(E) $[Co(H_2O)_3F_3]$

Choose the correct answer from the options given below : **[JEE Main (Jan.) 2025]**

(a) (B) and (D) only (b) (C) and (E) only
(c) (A), (B) and (D) only (d) (A), (C) and (E) only

B. Objective Questions

[One or more than one option(s) is/are correct]

1. The compound(s) that exhibit(s) geometrical isomerism is(are): **(IIT 2009)**

(a) $[Pt(en)Cl_2]$ (b) $[Pt(en)_2Cl_2]$ (c) $[Pt(en)_2Cl_2]Cl_2$ (d) $[Pt(NH_3)_2Cl_2]$

2. The pair(s) of coordination complexes/ions exhibiting the same kind of isomerism is (are) : **[JEE (Advanced) 2013]**

(a) $[Cr(NH_3)_5Cl]Cl_2$ and $[Cr(NH_3)_4Cl_2]Cl$
(b) $[Co(NH_3)_4Cl_2]^+$ and $[Pt(NH_3)_2(H_2O)Cl]^+$
(c) $[CoBr_2Cl_2]^{2-}$ and $[PtBr_2Cl_2]^{2-}$
(d) $[Pt(NH_3)_3(NO_3)]Cl$ and $[Pt(NH_3)_3Cl]Br$

3. Addition of excess aqueous ammonia to a pink coloured aqueous solution of $MCl_2 \cdot 6H_2O$ (*X*) and NH_4Cl gives an octahedral complex *Y* in the presence of air. In aqueous solution, complex *Y* behaves as 1:3 electrolyte. The reaction of *X* with excess HCl at room temperature results in the formation of a blue coloured complex *Z*. The calculated spin only magnetic moment of *X* and *Z* is 3.87 B.M., whereas it is zero for complex *Y*. Among the following options, which statement (s) is (are) correct? **[JEE (Advanced) 2017]**

(a) The hybridization of the central metal ion in Y is d^2sp^3.
(b) When X and Z are in equilibrium at 0° C, the colour of the solution is pink.
(c) Z is a tetrahedral complex.
(d) Addition of silver nitrate to Y gives only two equivalents of silver chloride.

4. The correct statement(s) regarding the binary transition metal carbonyl compounds is (are) (Atomic numbers: Fe = 26, Ni = 28): **[JEE (Advanced) 2018]**
(a) Total number of valence shell electrons at metal centre in $Fe(CO)_5$ or $Ni(CO)_4$ is 16.
(b) These are predominantly low spin in nature.
(c) Metal-carbon bond strengthens when the oxidation state of the metal is lowered.
(d) The carbonyl C—O bond weakens when the oxidation state of the metal is increased.

5. Choose the correct statement(s) among the following : **[JEE (Advanced) 2020]**
(a) $[FeCl_4]^-$ has tetrahedral geometry.
(b) $[Co(en)(NH_3)_2Cl_2]^+$ has 2 geometrical isomers.
(c) $[FeCl_4]^-$ has higher spin-only magnetic moment than $[Co(en)(NH_3)_2Cl_2]^+$.
(d) The cobalt ion in $[Co(en)(NH_3)_2Cl_2]^+$ has sp^3d^2 hybridization.

C. Matching Type Problems

1. Match the reactions in **Column I** with their properties listed in **Column II.** **(IIT 2007)**

	Column-I		Column-II
(a)	$[Co(NH_3)_4(H_2O)_2]Cl_2$	(p)	Geometrical isomers
(b)	$[Pt(NH_3)_2Cl_2]$	(q)	Paramagnetic
(c)	$[Co(H_2O)_5C]Cl$	(r)	Diamagnetic
(d)	$[Ni(H_2O)_6]Cl_2$	(s)	Metal ion with + 2 oxidation state

2. Match each co-ordination compound in List-I with an appropriate pair of characteristics from List-II and select the correct answer using the code given below the lists.
{en = $H_2NCH_2CH_2NH_2$; atomic number : Ti = 22, Cr = 24; Co = 27; Pt = /8}
[JEE (Advanced) 2014]

	List-I		List-II
(P)	$[Cr(NH_3)_4Cl_2]Cl$	1.	Paramagnetic and exhibits ionization isomerism
(Q)	$[Ti(H_2O)_5Cl](NO_3)_2$	2.	Diamagnetic and exhibits cis-trans isomerism
(R)	$[Pt(en)(NH_3)Cl]NO_3$	3.	Paramagnetic and exhibits cis-trans isomerism
(S)	$[Co(NH_3)_4(NO_3)_2]NO_3$	4.	Diamagnetic and exhibits ionization isomerism

Code :

	P	Q	R	S
(a)	4	2	3	1
(b)	3	1	4	2
(c)	2	1	3	4
(d)	1	3	4	2

3. Match List-I with List-II. **[JEE Main (Jan.) 2025]**

	List-I (Complex)		List-II (Hybridisation and Magnetic characters)
(A)	$[MnBr_4]^{2-}$	(i)	d^2sp^3 and diamagnetic
(B)	$[FeF_6]^{3-}$	(ii)	sp^3d^2 and paramagnetic
(C)	$[Co(C_2O_4)_3]^{3-}$	(iii)	sp^3 and diamagnetic
(D)	$[Ni(CO)_4]$	(iv)	sp^3 and paramagnetic

Choose the correct answer from the options given below :

(a) (A)-(III), (B)-(II), (C)-(I), (D)-(IV) (b) (A)-(III), (B)-(I), (C)-(II), (D)-(IV)
(c) (A)-(IV), (B)-(I), (C)-(II), (D)-(III) (d) (A)-(IV), (B)-(II), (C)-(I), (D)-(III)

D. Passage Based Problems

Passage-1

The coordination number of Ni^{2+} is 4.

$NiCl_2 + KCN\,(excess) \longrightarrow A$ (cyano complex)

$NiCl_2 + conc.\,HCl\,(excess) \longrightarrow B$ (chloro complex) **(IIT 2006)**

1. The IUPAC name of *A* and *B* are :

(a) potassium tetracyanonickelate (II), potassium tetrachloronickelate (II)
(b) tetracyanopotassiumnickelate (II), tetrachloropotassiumnickelate (II)
(c) tetracyanonickel (II), tetrachloronickel (II)
(d) potassium tetracyanonickel (II), potassium tetrachloronickel (II)

2. Predict the magnetic nature of *A* and *B*.

(a) Both are diamagnetic
(b) *A* is diamagnetic and *B* is paramagnetic with one unpaired electron.
(c) *A* is diamagnetic and *B* is paramagnetic with two unpaired electrons.
(d) Both are paramagnetic.

3. The hybridisation of *A* and *B* are :

(a) dsp^2, sp^3 (b) sp^3, sp^3
(c) dsp^2, dsp^2 (d) sp^3d^2, d^2sp^3

Passage-2

p-Amino-*N*, *N*-dimethylaniline is added to a strongly acidic solution of **X**. The resulting solution is treated with a few drops of aqueous solution of **Y** to yield blue coloration due to the formation of methylene blue. Treatment of the aqueous solution of **Y** with the reagent potassium hexacyanoferrate (II) leads to the formation of an intense blue precipitate. The precipitate dissolves on excess addition of the reagent. Similarly, treatment of the solution of **Y** with the solution of potassium hexacyanoferrate (III) leads to a brown coloration due to the formation of **Z**. **(IIT 2009)**

1. The compound **X** is :
(a) $NaNO_3$ (b) NaCl (c) Na_2SO_4 (d) Na_2S

2. The compound **Y** is :
(a) $MgCl_2$ (b) $FeCl_2$ (c) $FeCl_3$ (d) $ZnCl_2$

3. The compound **Z** is :
(a) $Mg_2[Fe(CN)_6]$ (b) $Fe[Fe(CN)_6]$ (c) $Fe_4[Fe(CN)_6]_3$ (d) $K_2Zn_3[Fe(CN)_6]_2$

E. Statement and Explanation Type Problems

Read the following questions and answer as per the direction given below :

(a) Statement 1 is true; Statement 2 is true; Statement 2 is a correct explanation of Statement 1.
(b) Statement 1 is true; Statement 2 is true; Statement 2 is not the correct explanation of Statement 1.
(c) Statement 1 is true; Statement 2 is false.
(d) Statement 1 is false; Statement 2 is true.

1. Statement 1 : $[Fe(H_2O)_5NO]SO_4$ is paramagnetic.
Statement 2 : The Fe in $[Fe(H_2O)_5NO]SO_4$ has three unpaired electrons. **(IIT 2008)**

2. Statement 1 : The geometrical isomer of the complex $[M(NH_3)_4Cl_2]$ are optically inactive.
Statement 2 : Both geometrical isomers of the complex $[M(NH_3)_4Cl_2]$ posses axis of symmetry. **(IIT 2008)**

F. Integer Type Problems

1. Total number of geometrical isomers for the complex $[RhCl(CO)(PPh_3)(NH_3)]$ is : **(IIT 2010)**

2. EDETA^{4-} is ethylenediaminetetraacetate ion. The total number of N — Co — O bond angles in $[Co(EDTA)]^{1-}$ complex ion is : **[JEE (Advanced) 2013]**

3. For the octahedral complexes of Fe^{3+} in SCN^- (thiocyanato-S) and in CN^- ligand environments, the difference between the spin-only magnetic moments in Bohr magnetons (When approximated to the nearest integer) is : [Atomic number of Fe = 26] **[JEE (Advanced) 2015]**

4. In dilute aqueous H_2SO_4, the complex diaquodioxalatoferrate(II) is oxidized by MnO_4^-. For this reaction, the ratio of the rate of change of $[H^+]$ to the rate of change of $[MnO_4^-]$ is :

[JEE (Advanced) 2015]

5. In the complex acetylbromidodicarbonylbis (triethylphosphine)iron(II), the number of Fe — C bond(s) is : **[JEE (Advanced) 2015]**

6. Among the complex ions,
$[Co(NH_2-CH_2-CH_2-NH_2)_2Cl_2]^+$, $[CrCl_2(C_2O_4)_2]^{3-}$, $[Fe(H_2O)_4(OH)_2]^+$, $[Fe(NH_3)_2(CN)_4]^-$, $[Co(NH_2-CH_2-CH_2-NH_2)_2(NH_3)Cl]^{2+}$ and $[Co(NH_3)_4(H_2O)Cl]^{2+}$, the number of complex ion(s) that show(s) *cis-trans* isomerism is :

[JEE (Advanced) 2015]

7. The number of geometric isomers possible for the complex $[CoL_2Cl_2]^-$ $(L = H_2NCH_2CH_2O^-)$ is : **[JEE (Advanced) 2016]**

8. Among the species given below, the total number of diamagnetic species is ....
H atom, NO_2 monomer, O_2^- (superoxide), dimeric sulphur in vapour phase, Mn_3O_4, $(NH_4)_2[FeCl_4]$, $(NH_4)_2[NiCl_4]$, K_2MnO_4, K_2CrO_4 **[JEE (Advanced) 2018]**

9. Total number of *cis* N — Mn — Cl bond angles (that is, Mn — N and Mn — Cl bonds in *cis* positions) present in a molecule of *cis*-$[Mn(en)_2Cl_2]$ complex is:
$(en = NH_2CH_2CH_2NH_2)$ **[JEE (Advanced) 2019]**

10. The volume (in mL) of 0.125 *M* $AgNO_3$ required to quantitatively precipitate chloride ions in 0.3 g of $[Co(NH_3)_6]Cl_3$ is ——— **[JEE Main (Jan.) 2020]**
$^{M}[Co(NH_3)_6Cl_3] = 267.46$ g/mol, $^{M}AgNO_3 = 169.87$ g/mol

11. Complexes (ML_5) of metals Ni and Fe have ideal square pyramidal and trigonal bipyramidal geometries, respectively. The sum of the 90°, 120° and 180° *L-M-L* angles in the two complexes is ______. **[JEE Main (Jan.) 2020]**

12. Considering that $\Delta_0 > P$, the magnetic moment (in BM) of $[Ru(H_2O)_6]^{2+}$ would be ______.

[JEE Main (Sept.) 2020]

13. The number of stereoisomers possible for $[Co(ox)_2(Br)(NH_3)]^{2-}$ is ______. [ox = oxalate]

[JEE Main (Feb.) 2021]

14. $[Ti(H_2O)_6]^{3+}$ absorbs light of wavelength 498 nm during a $d-d$ transition. The octahedral splitting energy for the above complex is ______ $\times 10^{-19}$ J. (Round off to the nearest integer).
$h = 6.626 \times 10^{-34}$ Js; $c = 3 \times 10^8$ ms^{-1} **[JEE Main (March) 2021]**

15. In the ground state of atomic Fe($Z = 26$), the spin only- magnetic moment is .... $\times 10^{-1}$ BM. (Round off to the nearest integer).
[Given: $\sqrt{3} = 1.73, \sqrt{2} = 1.41$] **[JEE Main (March) 2021]**

16. On complete reaction of $FeCl_3$ with oxalic acid in aqueous solution containing KOH, resulted in the formation of product *A*. The secondary valency of Fe in the product *A* is.....
(Round off to the nearest integer). **[JEE Main (March) 2021]**

17. The total number of unpaired electrons present in the complex $K_3[Cr(oxalate)_3]$ is.......

[JEE Main (March) 2021]

18. Sum of oxidation state (magnitude) and coordination number of cobalt in $Na[Co(bpy)Cl_4]$ is ______.

(Given bpy = 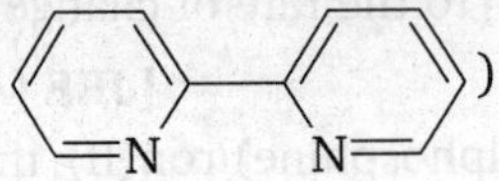) **[JEE Main (July) 2022 (II)]**

19. Total number of relatively more stable isomer(s) possible for octahedral complex $[Cu(en)_2(SCN)_2]$ will be ______. **[JEE Main (July) 2022 (I)]**

20. The conductivity of a solution of complex with formula $CoCl_3(NH_3)_4$ corresponds to 1 : 1 electrolyte, then the primary valency of central metal ion is ______.

[JEE Main (July) 2022 (I)]

21. Spin only magnetic moment of $[MnBr_6]^{4-}$ is ______ B.M.

(Round off to the closed integer) **[JEE Main (June) 2022 (II)]**

22. The spin only magnetic moment of the complex present in Fehling's reagent is ______ B.M. (Nearest integer). **[JEE Main (July) 2022 (II)]**

23. The ratio of spin-only magnetic moment values $\mu_{eff}[Cr(CN)_6]^{3-}/\mu_{eff}[Cr(H_2O)_6]^{3+}$ is ______.

[JEE Main (April) 2023 (I)]

24. In potassium ferrocyanide, there are ______ pairs of electrons in the $t2g$ set of orbitals :

[JEE Main (April) 2023 (I)]

25. The observed magnetic moment of the complex $[Mn(\underline{N}CS)_6]^{x-}$ is 6.06 BM. The numerical value of x is ______: **[JEE Main (April) 2023 (II)]**

26. If the CFSE of $[Ti(H_2O)_6]^{3-}$ is –96.0 kJ/mol, this complex will absorb maximum at wavelength ______ nm. (Nearest integer)

Assume Planck's constant $(h) = 6.4 \times 10^{-34}$ Js, Speed of light $(c) = 3.0 \times 10^8$ m/s and Avogadro's constant $(N_A) = 6 \times 10^{23}$/mol. **[JEE Main (Jan.) 2023 (II)]**

27. The oxidation number of iron in the compound formed during brown ring test for NO_3^- ion is ______. **[JEE Main (Jan.) 2024]**

28. The complex of Ni^{2+} ion and dimethyl glyoxime contains .... number of hydrogen (H) atoms. **[JEE Main (Jan.) 2025]**

29. Among the following cations, the number of cations which will give characteristic precipitate in their identification tests with $K_4[Fe(CN)_6]$ is :

Cu^{2+}, Fe^{3+}, Ba^{2+}, Ca^{2+}, NH_4^+, Mg^{2+}, Zn^{2+} **[JEE Main (Jan.) 2025]**

30. The spin only magnetic moment (μ) value (B.M.) of the compound with strongest oxidising power among Mn_2O_3, TiO and VO is .... B.M. (Nearest integer).

[JEE Main (Jan.) 2025]

31. Total number of molecules/species from following which will be paramagnetic is ....

O_2, O_2^+, O_2^-, NO, NO_2, CO, $K_2[NiCl_4]$, $[Co(NH_3)_6]Cl_3$, $K_2[Ni(CN)_4]$

[JEE Main (Jan.) 2025]

32. Consider the following low-spin complexes

$K_3[Co(NO_2)_6]$, $K_4[Fe(CN)_6]$, $K_3[Fe(CN)_6]$, $Cu_2[Fe(CN)_6]$ and $Zn_2[Fe(CN)_6]$.

The sum of the spin-only magnetic moment values of complexes having yellow colour is .... B.M. (answer is nearest integer) **[JEE Main (Jan.) 2025]**

Answers

[A] Objective Questions (Only one option is correct)

1. (c) **2.** (b) **3.** (b) **4.** (b) **5.** (c) **6.** (c) **7.** (a) **8.** (d) **9.** (c) **10.** (b)
11. (b) **12.** (b) **13.** (c) **14.** (d) **15.** (a) **16.** (b) **17.** (b) **18.** (b) **19.** (a) **20.** (b)
21. (d) **22.** (a) **23.** (b) **24.** (d) **25.** (d) **26.** (d) **27.** (b) **28.** (b) **29.** (a) **30.** (b)
31. (b) **32.** (d) **33.** (a) **34.** (b) **35.** (a) **36.** (d) **37.** (a) **38.** (b) **39.** (d) **40.** (b)
41. (b) **42.** (b) **43.** (c) **44.** (a) **45.** (b) **46.** (c) **47.** (a) **48.** (c) **49.** (a) **50.** (c)
51. (a) **52.** (c) **53.** (a) **54.** (a) **55.** (a) **56.** (d) **57.** (c) **58.** (a) **59.** (a) **60.** (b)
61. (b) **62.** (d) **63.** (d) **64.** (a) **65.** (a) **66.** (c) **67.** (a) **68.** (d) **69.** (b) **70.** (b)
71. (c) **72.** (c) **73.** (a) **74.** (a) **75.** (c) **76.** (a) **77.** (b) **78.** (c) **79.** (a) **80.** (c)
81. (b) **82.** (a)

[B] Objective Questions (One or more than one option(s) is/are correct)

1. (c, d) **2.** (b, d) **3.** (a, b, c) **4.** (b, c) **5.** (a, c)

[C] Matching Type Problems

1. $a \rightarrow p, q, s$; $b \rightarrow p, r, s$; $c \rightarrow q, s$; $d \rightarrow q, s$
2. $P \rightarrow 3$, $Q \rightarrow 1$; $R \rightarrow 4$; $S \rightarrow 2$
3. (A)-(IV), (B)-(II), (C)-(I), (D)-(III)

[D] Passage Based Problem

Passage-1

1. (a) **2.** (c) **3.** (a)

Passage-2

1. (d) **2.** (c) **3.** (b)

[E] Statement and Explanation Type Problems

1. (a) **2.** (a)

[F] Integer Type Problems

1. (3)	**2.** (8)	**3.** (4)	**4.** (8)	**5.** (3)
6. (6)	**7.** (5)	**8.** (1)	**9.** (6)	**10.** (26.92)
11. (20)	**12.** (0)	**13.** (3)	**14.** (4)	**15.** (49)
16. (6)	**17.** (3)	**18.** (9)	**19.** (3)	**20.** (3)
21. (6)	**22.** (2)	**23.** (1)	**24.** (3)	**25.** (4)
26. (480)	**27.** (1)	**28.** (14)	**29.** (3)	**30.** (5)
31. (6)	**32.** (0)			

Hints and Solutions

A. Objective Questions

[Only one option is correct]

1. (c) It is a fact.

2. (b) $\underset{(A)}{[Cr(H_2O)_4Cl(NO_2)]Cl} \rightleftharpoons [Cr(H_2O)_4 \cdot Cl(NO_2)]^+ + Cl^-$

$\underset{(B)}{[Cr(H_2O)_4Cl_2]NO_2} \rightleftharpoons [Cr(H_2O)_4 \cdot Cl_2]^+ + NO_2^-$

A and *B* have same M.F. but furnish different ions in solution.

3. (b) $\mu = \sqrt{n(n+2)}$

or $2.83 = \sqrt{n(n+2)} \Rightarrow n = 2$

i.e., species with two unpaired electrons will show $\mu = 2.83$ B.M.

In $[NiCl_4]^{2-}$, O.S. of Ni = +2

$Ni(28) = 3d^8 4s^2$

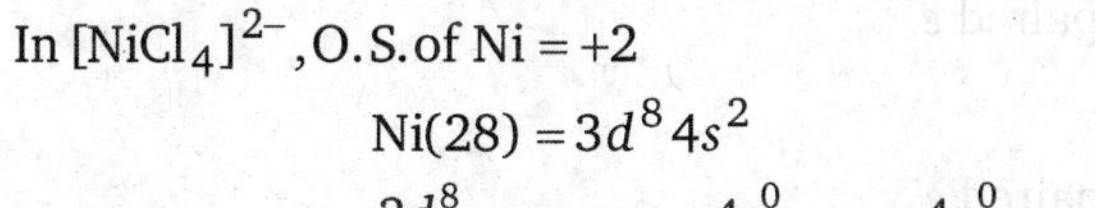

Cl^- being weak ligand can not pair up 3*d*-electrons

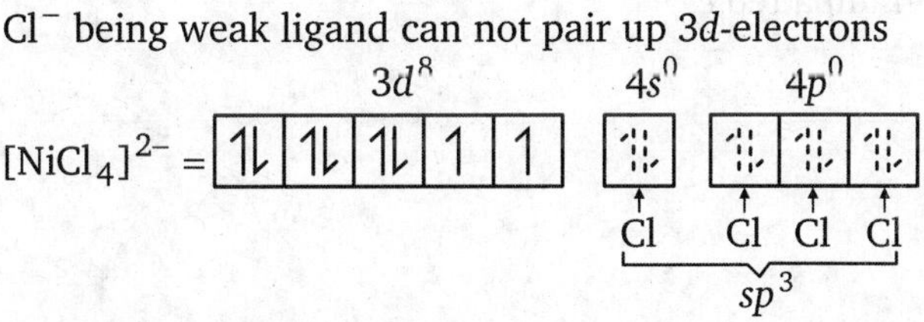

4. (b) $[NiCl_4]^{2-} \rightarrow$ Tetrahedral, $[Ni(CN)_4]^{2-} \rightarrow$ Square Planar

$[Ni(H_2O)_6]^{2+} \rightarrow$ Octahedral

5. (c) Following compounds are diamagnetic.

L: $[Co(NH_3)_6]Cl_3$ M: $Na_3[Co(Ox)_3]$

O: $K_2[Pt(CN)_4]$ P: $[Zn(H_2O)_6](NO_3)_2$

6. (c) Electronic configuration of Ni: $[Ar]3d^8 4s^2$

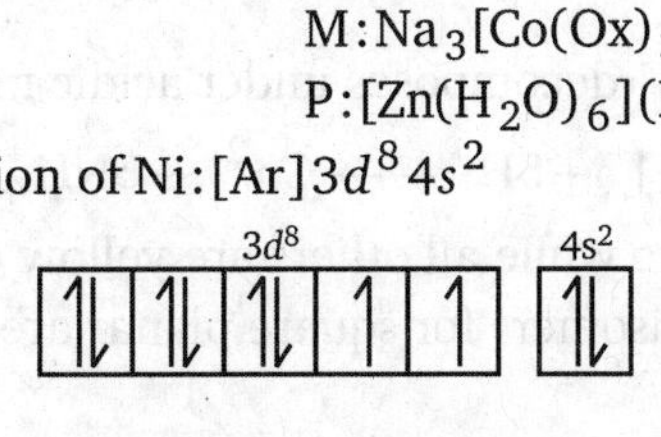

$Ni^{2+}: [Ar]3d^8 4s^0$

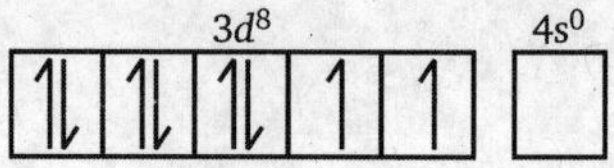

Paramagnetic behaviour is possible when pairing does not take place. *i.e.*, $3d$ will not participate in bonding and hybridisation will be sp^3 including $4s$ and $4p$, thus structure is tetrahedral.

When pairing takes places (in presence of strong ligand field),

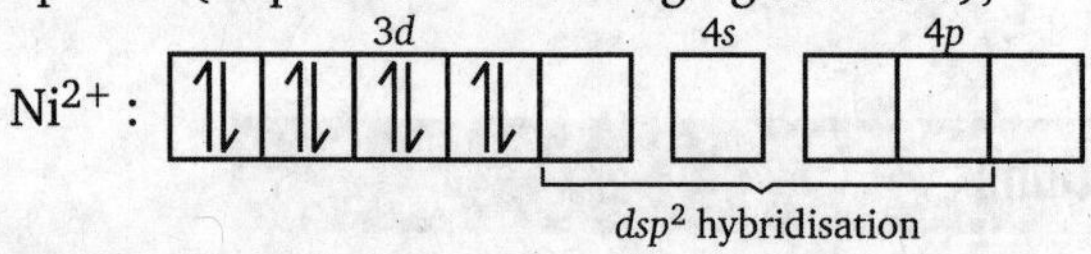

dsp^2 hybrid orbitals arrange in square planar structure.

Ni^{2+}: paramagnetic, sp^3, tetrahedral

Ni^{2+}: diamagnetic, dsp^2, square planar

7. (a) The colour of aqueous solution of $CuSO_4$ is blue green. Thus it absorbs orange-red colour and exhibit the complementary colour.

8. (d) Diamminetetraaquacobalt(III) chloride

9. (c) $[Co(NH_3)_3Cl_3]$ will not exhibit optical isomerism due to presence of plane of symmetry.

10. (b) $Q < R < P$

$P = [FeF_6]^{3-}$

$Fe^{3+} = [Ar]\,4s^0 3d^5$ 5 unpaired e^-

$Q = [V(H_2O)_6]^{2+}$

$V^{2+} = [Ar]\,4s^0 3d^3$ 3 unpaired e^-

$R = [Fe(H_2O)_6]^{2+}$

$Fe^{2+} = [Ar]\,4s^0 3d^6$ 4 unpaired e^-

11. (b)

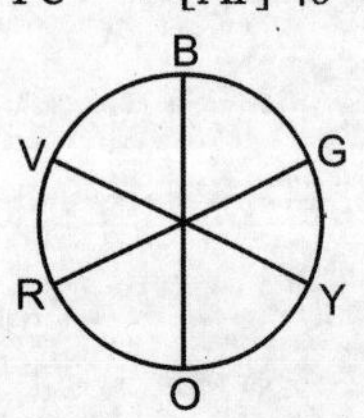

The energy of red light is less than that of violet light.

So energy order is Red < Yellow < Green < Blue.

The complex absorbs lower energy light lower will be its strength. So order of ligand strength is $L_1 < L_3 < L_2 < L_4$.

12. (b) The complex $[CoCl(NH_3)_5]^+$ decomposes under acidic medium, so

$$[CoCl(NH_3)_5]^+ + 5H^+ \longrightarrow Co^{2+} + 5NH_4^+ + Cl^-.$$

13. (c) $Zn_2[Fe(CN)_6]$ is bluish white while all others are yellow coloured.

14. (d) The number of geometrical isomers for square planar are 3.

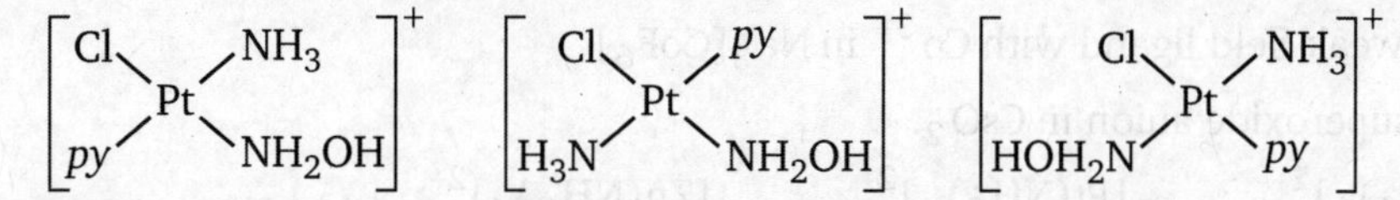

15. (a) The deep purple colour of $KMnO_4$ is not due to d - d transitions but due to charge transfer from O to Mn (*i.e.*, $L \rightarrow M$), which reduces the oxidation state of Mn from +7 to +6 momentarily.

16. (b) $[Cr(H_2O)_6]^{2+}$

$$Cr^{2+} \longrightarrow [Ar]3d^4 4s^0$$

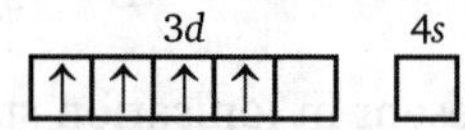

4 unpaired electrons

$[CoCl_4]^{2-}$

$$Co^{2+} \longrightarrow [Ar]3d^7 4s^0$$

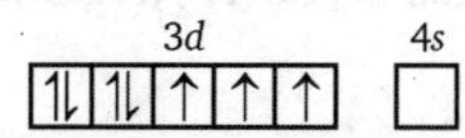

3 unpaired electrons

$[Fe(H_2O)_6]^{2+}$

$$Fe^{2+} \longrightarrow [Ar]3d^6 4s^0$$

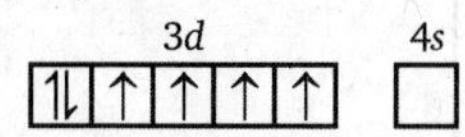

4 unpaired electrons

$[Mn(H_2O)_6]^{2+}$

$$Mn^{2+} \longrightarrow [Ar]3d^5 4s^0$$

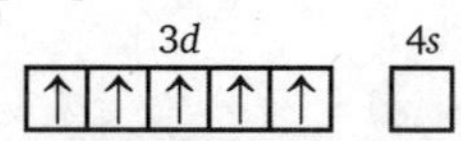

5 unpaired electrons

Hence, $[Cr(H_2O)_6]^{2+}$ and $[Fe(H_2O)_6]^{2+}$ have same number of unpaired electrons *i.e.*, same magnetic moment.

17. (b) $[Co(NH_3)_3Cl_3]$ has two geometrical isomers but both are optically inactive due to plane of symmetry.

cis $[Co(en)_2Cl_2]Cl$ is optically active.

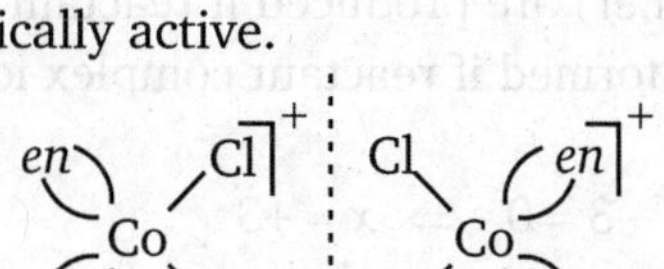

Non-superimposable

trans $[Co(en)_2Cl_2]Cl$ is optically inactive due to plane of symmetry.

$[Co(NH_3)_4Cl_2]Cl$ has two geometrical isomers but both are optically inactive due to plane of symmetry.

18. (b) $[NiCl_4]^{2-}$, $Na_3[CoF_6]$ and CsO_2 are paramagnetic.

Cl^- is a weak field ligand with Ni^{2+} in $[NiCl_4]^{2-}$

F^- is a weak field ligand with Co^{3+} in $Na_3[CoF_6]$

O_2^{2-} is superoxide anion in CsO_2.

19. (a) $[Ni(NH_3)_6]^{2+}$, sp^3d^2 Octahedral; $[Pt(NH_3)_4]^{2+}$, dsp^2 Square planar; $[Zn(NH_3)_4]^{2+}$ sp^3 Tetrahedral

20. (b) Number of moles of complex $= \dfrac{M \times V(\text{mL})}{1000} = \dfrac{0.1 \times 100}{1000} = 0.01$

Moles of ions precipitated with excess of $AgNO_3 = \dfrac{1.2 \times 10^{22}}{6.022 \times 10^{23}} = 0.01992 \approx 0.02$

Now, number of Cl^- ions present in ionisation sphere

$$= \frac{\text{Moles of ions precipitated with excess } AgNO_3}{\text{Moles of complex}} = \frac{0.02}{0.01} = 2$$

Hence, the formula of complex is $[Co(H_2O)_5Cl]Cl_2 \cdot H_2O$.

21. (d) Case-I

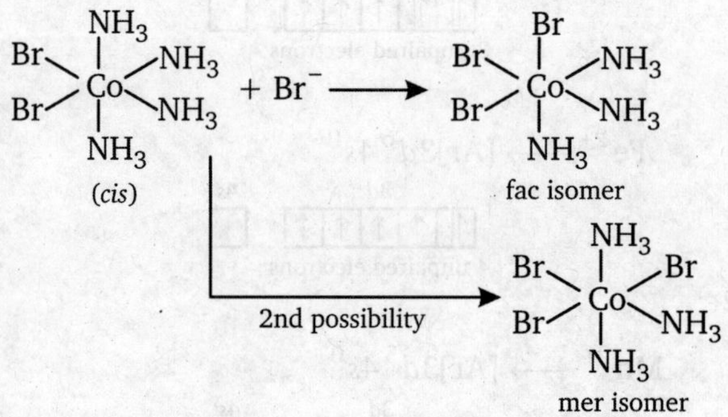

Case-II

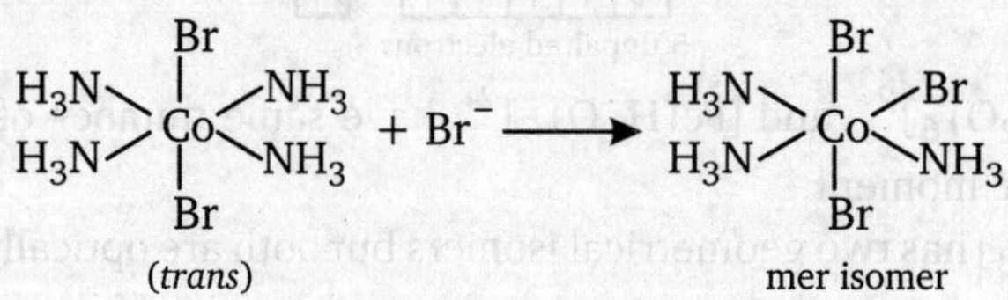

$\therefore$ Two isomers (fac and mer) are produced if reactant complex ion is a *cis* isomer.
Only one isomer (fac) is formed if reactant complex ion is a *trans* isomer.

22. (a) $[Cr(H_2O)_6]Cl_3$

$$x + 0 - 3 = 0 \Rightarrow x = +3$$

$[Cr(C_6H_6)_2]$

$$x + 0 = 0 \Rightarrow x = 0$$

$K_2[Cr(CN)_2(O)_2(O_2)(NH_3)]$

$$2 + x - 2 - 4 - 2 + 0 = 0 \Rightarrow x = +6$$

23. (b)

24. (d) $[V(CN)_6]^{4-}$ *i.e.*, $V^{2+} \Rightarrow 3d^3$

Magnetic moment $(\mu) = \sqrt{3(3+2)} = \sqrt{15} = 3.87$ B.M.

$[Fe(CN)_6]^{4-}$ *i.e.*, $Fe^{2+} \Rightarrow 3d^6$

Unpaired electrons = 0

$$\mu = 0$$

$[Ru(NH_3)_6]^{3+}$ *i.e.*, $Ru^{3+} \Rightarrow 4d^5$

Unpaired electron = 1

$$\mu = \sqrt{1(1+2)} = \sqrt{3} = 1.73$$

$[Cr(NH_3)_6]^{2+}$ *i.e.*, $Cr^{2+} \Rightarrow 3d^4$

Unpaired electrons = 2

$$\mu = \sqrt{2(2+2)} = \sqrt{6} = 2.45$$

Thus, the correct order is,

$$V^{2+} > Cr^{2+} > Ru^{3+} > Fe^{2+}$$

25. (d)

26. (d) Cationic species is $[Fe(H_2O)_6]^{2+}$.

Anionic species is $[Fe(CN)_6]^{4-}$.

In $[Fe(CN)_6]^{4-}$, $Fe^{2+} : 3d^6$

As CN^- is a strong field ligand, pairing of electrons takes place, therefore magnetic moment will be zero.

In $[Fe(H_2O)_6]^{2+}$, $Fe^{2+} : 3d^6$

As H_2O is a weak field ligand, four unpaired electrons are available, therefore magnetic moment is

$$\mu = \sqrt{n(n+2)} \quad (n = \text{no. of unpaired electrons})$$

$$\mu = \sqrt{4(4+2)} = \sqrt{24} = 4.89 \approx 4.9 \text{ BM}$$

Note : Charges on coordination complexes are not given in the question.

27. (b)

28. (b) $Cr(H_2O)_6Cl_n \xrightarrow{AgNO_3} AgCl \downarrow$

Complex *X*

$\mu = 3.83$ B.M. $\Rightarrow n = 3$ = No. of unpaired electrons.

- Since, charge on H_2O is zero, so Cr^{+n} will be having 3 unpaired electrons.
- Using the given options, *n* could be + 4 or +3

Cr^{3+}	Cr^{4+}
$t_{2g}^3 e_g^0$	$t_{2g}^2 e_g^0$
Unpaired electron	Unpaired electron
3	2
✓	✗

$Cr(H_2O)_6Cl_3$

(I) $[Cr(H_2O)_4Cl_2]Cl \cdot 2H_2O$ = ✓ name :

tetraaquadichlorido chromium (III) chloride dihydrate

(I) $[Cr(H_2O)_5Cl]Cl_2 \cdot H_2O$
(II) $[Cr(H_2O)_6]Cl_3$
] Does not show geometrical isomer

29. (a) Treating complex with conc. H_2SO_4, it removes water outside co-ordination sphere.
Let number of water molecules in complex is x.

$$\% \text{ mass of water} = \left(\frac{18 \times x}{12 + 6 \times 16 + 35 \times 3 + 52}\right) \times 100 = 13.5 \Rightarrow x = 2$$

So, complex will be $[Cr(H_2O)_4Cl_2]Cl \cdot 2H_2O$

30. (b) $[Pt(NH_3)_2Cl(NH_2CH_3)]Cl$
Diamminechlorido(methanamine)platinum(II) chloride
Alphabetical order
❖ Platinum → ium for cationic co-ordination sphere.

31. (b) Complex of type $[MA_3B_3]$ show F

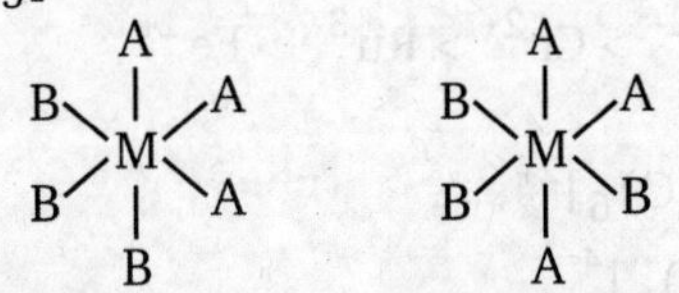

fac isomer mer isomer

Option (b) $[Co(NH_3)_3(NO_2)_3]$ will show geometrical isomerism.

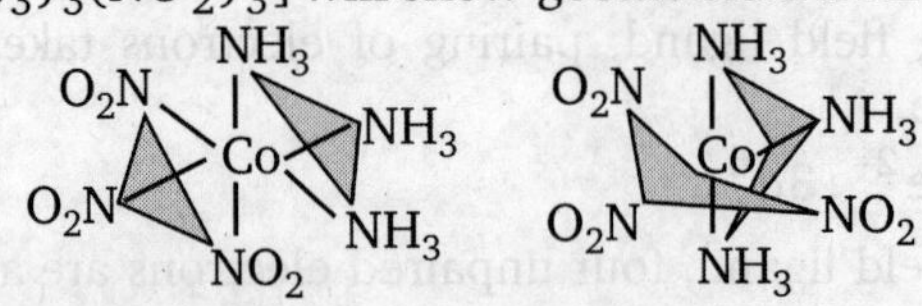

32. (d) $[Ni(NH_3)_2Cl_2]$ sp^3-hybridisation so no isomerism

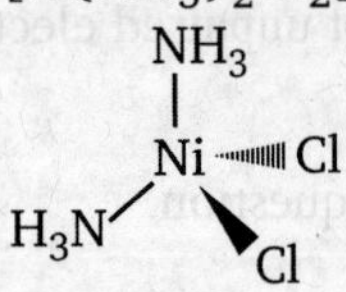

Plane of symmetry is present

33. (a) *cis* - $[CrCl_2(ox)_2]^{3-}$ (ox = oxalate)

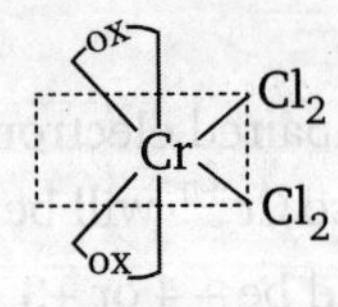

cis optically active compound.
❖ No plane of symmetry
❖ No centre of symmetry

34. (b)

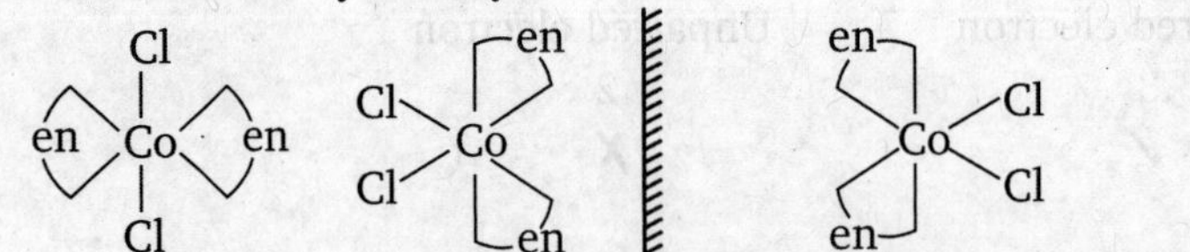

trans Plane of symmetry | *cis* | No plane of symmetry or centre of symmetry optically active

35. (a) $[Ni(CN)_4]^{2-}$

$$Ni^{2+} 1s^2 2s^2 2p^6 3s^2 3p^6 3d^8 4s^0$$

3d

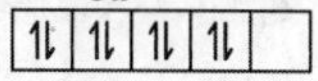

Hybridisation dsp^2

(i) Because all ligands are strong field ligands.

(ii) One electrons is shifting.

(iii) Inner orbital is being formed.

36. (d) (A) Octahedral Co(III) complexes with strong field ligand have low magnetic moment due to formation of lower spin complex.

(B) $\Delta_o < P$ so Co(III) $t_{2g}^4 e_g^2$

(C) en is stronger field ligand compare to fluoride anion (F^-)

$$\Delta_o(\text{en}) > \Delta_o(F^-)$$

$$\lambda_{en} < \lambda_{F^-}$$

(D) $\Delta_t = \Delta_o \times \frac{4}{9} = 18000 \times \frac{4}{9} = 8000\ \text{cm}^{-1}$

Statement (A) and (D) are incorrect.

37. (a) $[Cr(H_2O)_6]Br_2$ Cr^{2+} in presence of weak field ligand.

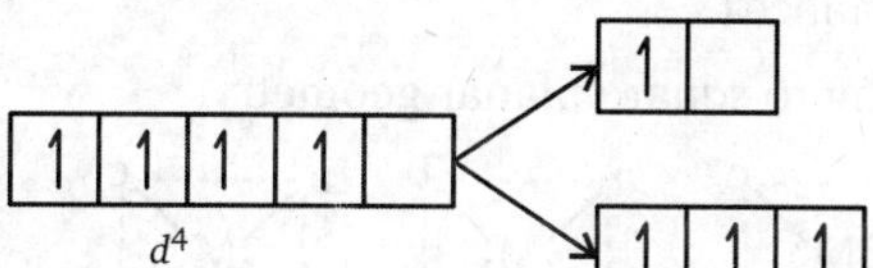

$n = 4$

$Na_4[Fe(CN)_6]$ Fe^{2+} in presence of strong field ligand.

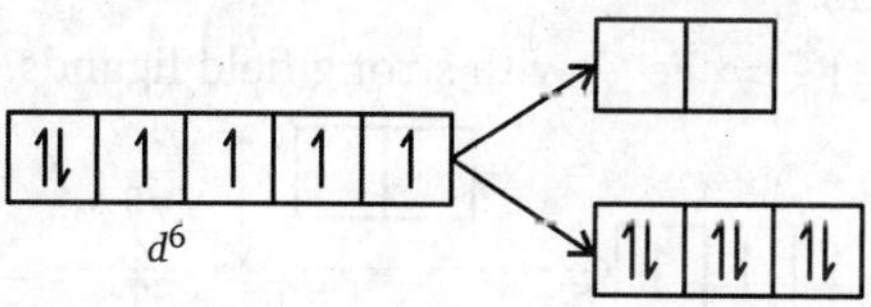

$n = 0$

$Na_3[Fe(C_2O_4)_3] (\Delta_0 > P)$ Fe^{3+}

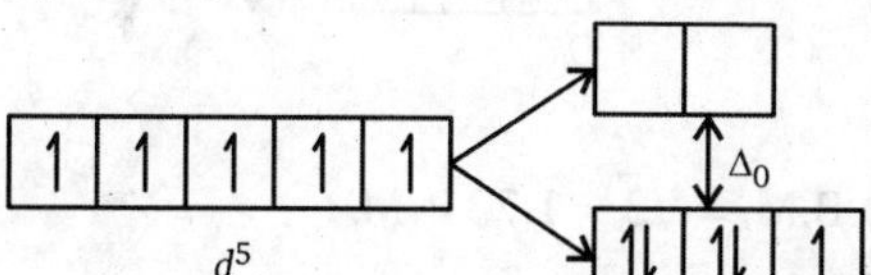

$\Delta_0 > P$
So pairing will start first

$n = 1$

$(Et_4N)_2[CoCl_4]$

Co^{2+} with halogen —a weak field ligand

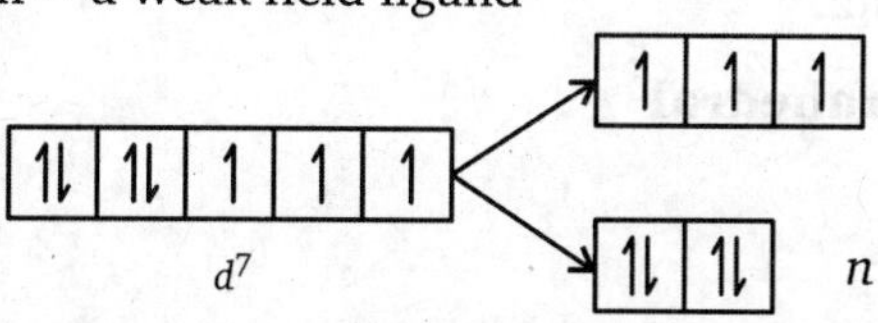

$n = 3$

(I) > (IV) > (III) > (II)

38. (b) (a) $[Fe(en)(bpy)(NH_3)_2]^{2+}$ Fe^{2+} $t_{2g}^{6}\ e_g^{0}$

(b) $[Co(ox)_2(OH)_2]^-$ Co^{5+} $t_{2g}^{2,1,1}\ e_g^{0,0}$

(c) $[Pd(gly)_2]$ Pd^{2+} $t_{2g}^{2,2,2}\ e_g^{2,0}$

dsp^2 hybridisation

(d) $[Ti(NH_3)_6]^{3+}$ Ti^{3+} $t_{2g}^{1}\ e_g^{0}$

Option (b) will have maximum paramagnetic behaviour.

39. (d) $\mu = 5.9$ BM

Number of unpaired electrons = 5

$[MnBr_4]^{2-}$ $Mn_2^+ = [Ar]3d^5 4s^0$

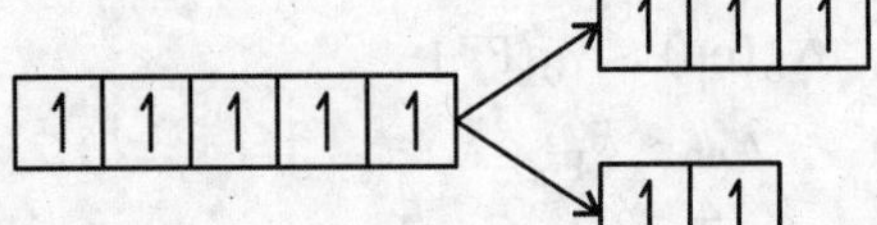

$Br^{\ominus}$ is a weak field ligand.

$n = 5$ $\mu = \sqrt{5(5+2)}$ BM = 5.9 BM

40. (b) $[Pd(F)(Cl)(Br)(I)]^{2-} = [Mabcd]$

dsp^2 hybridisation leading to square planar geometry.

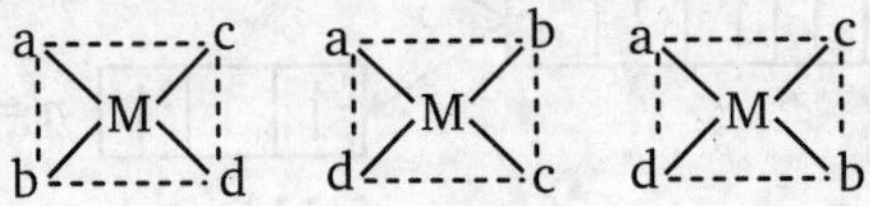

Three geometrical isomers.

$[Fe(CN)_6]^{3-6} = [Fe(CN)_6]^{3-} \Rightarrow Fe^{3+}$ with strong field ligands.

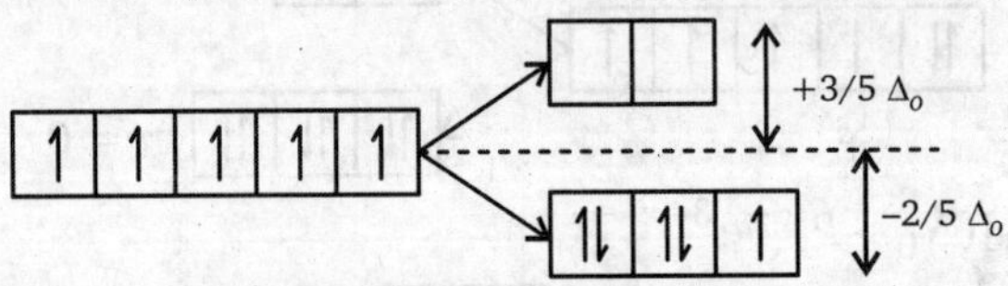

$$CFSE = -\frac{2}{5} \times 5\Delta_o = -2\Delta_o$$

$n = 1$; $\mu = \sqrt{n(n+2)}$ B.M. $= \sqrt{3} = 1.73$ B.M.

41. (b) Mn(II) complex: Octahedral

Mn (II) $t_{2g}^{1,1}\ e_g^{1}$ high spin

Mn (II) $t_{2g}^{2,1,1}\ e_g^{0}$ low spin

Ni(II) complex: Tetrahedral

Ni (II) $t_{2g}^{2,2,2}\ e_g^{1,1}$

Ni (II) $t_{2g}^{2,2,2}\ e_g^{2,0}$

dsp^2-hybridisation with strong carbonyl ligands.

Mn (II) (Mn^{2+}) is light pink colour in solution.

42. (b) d^6 metal ion (M^{2+}) form a complex with aqua ligands.

Spin only magnetic moment $= \sqrt{n(n+2)}$ B.M $= 4.90$ B.M.

$$n = 4$$

Octahedral splitting

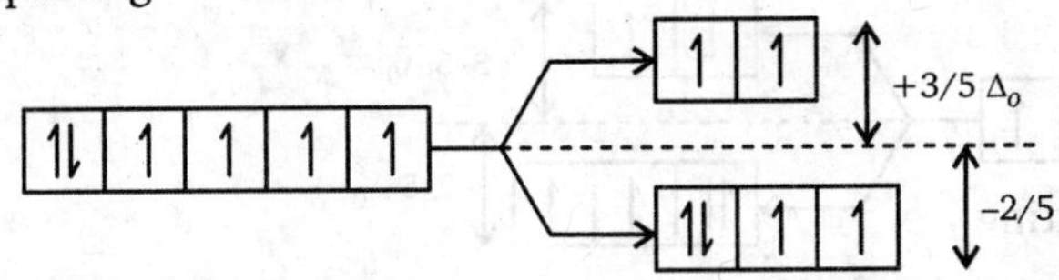

$$\text{CFSE} = -2/5\,\Delta_o \times 4 + \frac{3}{5} \times 2\Delta_o + P = \frac{-2}{6}\Delta_o + P$$

Tetrahedral splitting

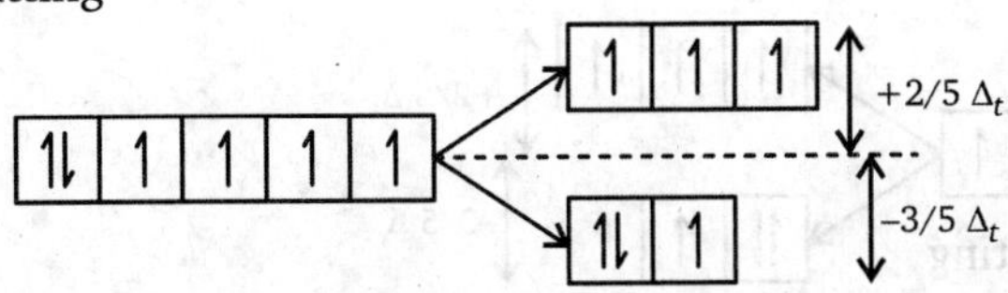

$$= -3/5 \times 3\Delta_t + 2/5\,\Delta_t \times 3 + P$$
$$= -0.6\Delta_t + P$$

Closest answer is (b) tetrahedral with CFSE $= -0.6\Delta_t + P$

43. (c)

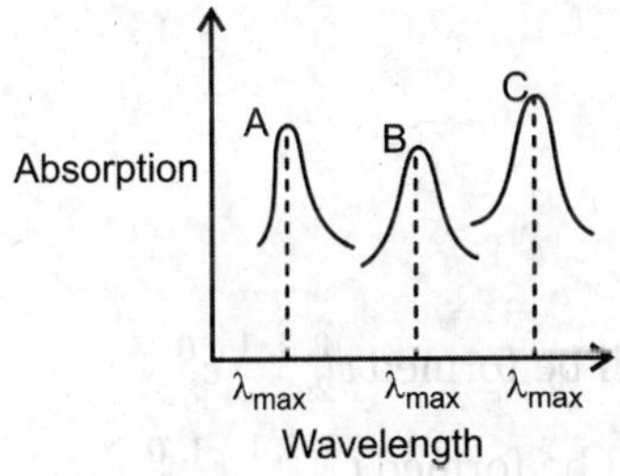

$[M(CNS)_6]^{(-6+n)}$ $[MF_6]^{(-6+n)}$ $[M(NH_3)_6]^{n+}$

Strength of ligand $NH_3 > NCS^- < F^{\ominus}$

CFSE order iii > i > ii

$$(\text{Absorption})\ \lambda_{max} \propto \frac{1}{\text{CFSE}}$$

λ_{max} order ii > i > iii

Order given in graph $C > B > A$

$$\text{i} = B;\ \text{ii} = C;\ \text{iii} = A$$

44. (a) $[Ru(en)_3]Cl_2$ | $[Fe(H_2O)_6]Cl_2$

$Ru = [Kr]4d^6\,5s^2$ | $Fe^{2+} = [Ar]3d^6 5s^0$

$Ru^{2+} = [Kr]\,4d^6 5s^0$

$t_{2g}^{2,2,2} e_g^{0,0}$ | $t_{2g}^{2,1,1}\ e_g^{1,1}$

- Inner orbital complex will be formed.

45. (b) CFSE of $[CoF_3(H_2O)_3)](\Delta_0 < P)$

Co^{3+}: $[Ar]3d^6 4s^0$

$t_{2g}^4\ e_g^2$

$CFSE = 4 \times (-2/5\,\Delta_0) + 2(2/5\,\Delta_0) + P = -0.4\Delta_0 + P$

Correct answer should be (b).

46. (c)

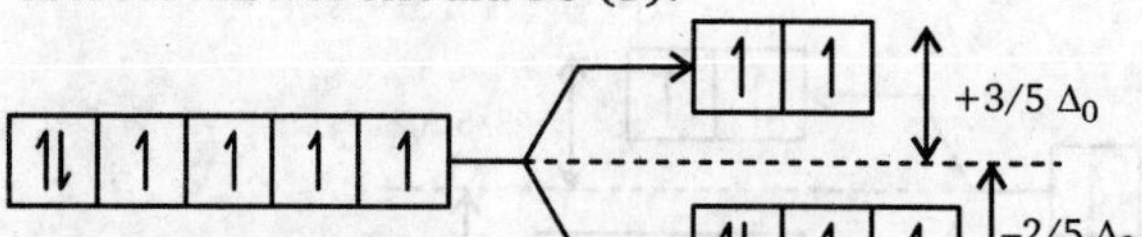

Octahedral splitting

High spin complex.

$$CFSE = -\frac{2}{5} \times 4(\Delta_0) + \frac{3}{5} \times 2\Delta_0$$

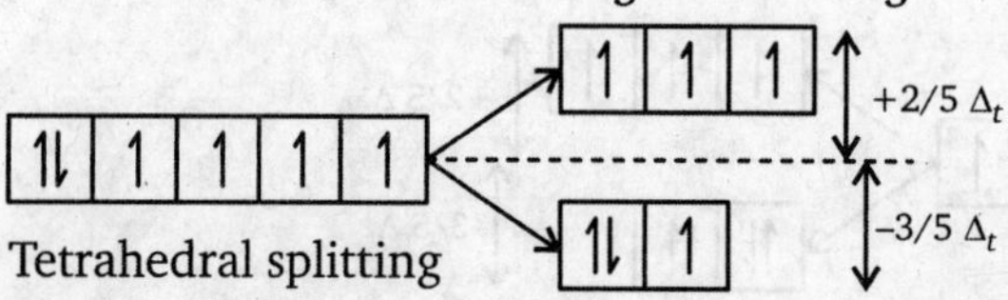

Tetrahedral splitting

High spin complex.

$$CFSE = -\frac{3}{5}(3)\Delta_t + \frac{2}{5} \times 3\Delta_t$$

$$CFSE\ \Delta_0 = -\frac{2}{5}\Delta_0 = -0.4\,\Delta_0$$

$$CFSE\ \Delta_t = -\frac{3}{5}\Delta_t = -0.6\,\Delta_t$$

47. (a) d^4 metal ion in an octahedral field.

- Strong field ligand $\Delta_0 > P$, low spin complex will be formed $t_{2g}^{2,1,1}\ e_g^0$
- Weak field ligand $\Delta_0 < P$, high spin complex will be formed $t_{2g}^{1,1,1}\ e_g^{1,0}$

48. (c) $Ni^{2+} + 2dmg^- \longrightarrow [Ni(dmg)_2]$

Rosy red complex (ppt.)

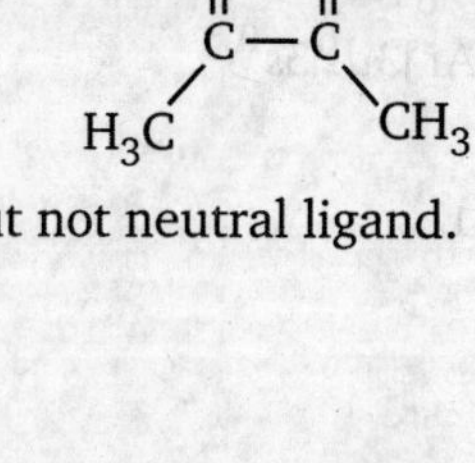

DMG is bidentate ligand but not neutral ligand.

49. (a) $[\mathbf{FeF_6}]^{3-}$ $Fe^{3+}[Ar]3d^5 4s^0$

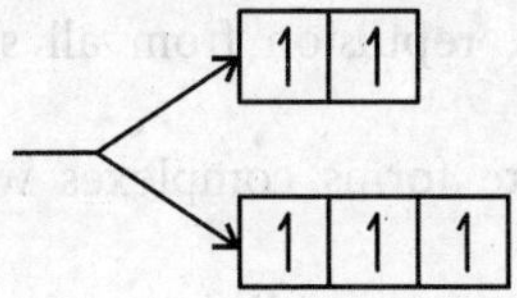

$\mu = \sqrt{35}$

$[\mathbf{Co(NH_3)_6}]^{3+}$ $Co^{3+}[Ar]3d^6 4s^0$

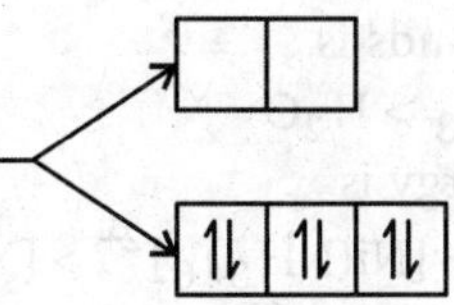

$\mu = 0$

$[\mathbf{NiCl_4}]^{2-}$ $Ni^{2+}[Ar]3d^8 4s^0$

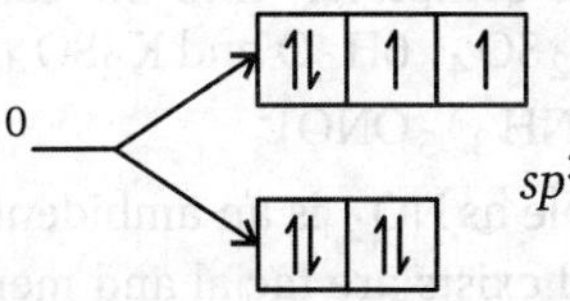

$\mu = \sqrt{8}$

$[\mathbf{Cu(NH_3)_4}]^{2+}$ $Cu^{2+}[Ar]3d^9 4s^0$

1 unpaired e^- $\mu = \sqrt{3}$

dsp^2

Correct order is i > iii > iv > ii

50. (c) $[Mn(CN)_6]^{4-}$ and $[Fe(CN)_6]^{3-}$

Mn^{2+} Fe^{3+}

$[Ar]3d^5 4s^0$ $[Ar]3d^5 4s^0$

Sincc, $CN^{\ominus}$ is a strong field ligand and pairing will start and it will lead to formation of inner orbital complex.

Both will show same configuration.

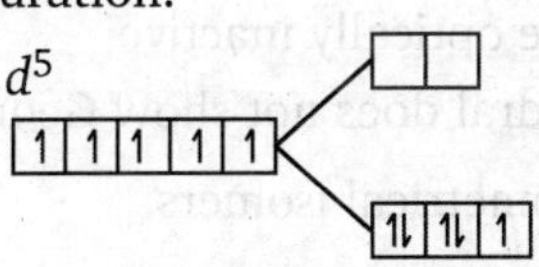

d^2sp^3-hybridisation.

Due to unpaired electron it will be paramagnetic.

51. (a) Only the counter ions are replaced easily by other reagents,

$$[Co(NH_3)_5Br]SO_4 + BaCl_2 \longrightarrow [Co(NH_3)_5Br]Cl_2 + \underset{\text{White ppt}}{BaSO_4\downarrow}$$

It does not have chloride ion as counter ion so it will not give white ppt with $AgNO_3$.

52. (c) $AgCl + 2NH_3 \longrightarrow \underset{\text{soluble}}{[Ag(NH_3)_2]^+Cl^-}$

53. (a) Strong Jahn-Teller distortion is observed in case of unsymmetrical filling of e_g orbitals.

When the ligands are same, repulsion from all side are symmetrical and maximum distortion is observed.

54. (a) Metals of low oxidation state forms complexes with ligands which have π-accepting character.

55. (a) Energy of absorption depends on the splitting of d-orbital in crystal field splitting.
A strong field ligand have high crystal field splitting.
Order of strength of ligands is

$$en > NH_3 > H_2O$$

Hence, the order of energy is

$$\underset{(A)}{[Ni(en)_3]^{2+}} > \underset{(B)}{[Ni(NH_3)_6]^{2+}} > \underset{(C)}{[Ni(H_2O)_6]^{2+}}$$

56. (d) The double salts dissociates completely into the constituting ions on dissolution. This is possible for $FeSO_4 \cdot (NH_4)_2SO_4 \cdot 6H_2O$ and $K_2SO_4 \cdot Al_2(SO_4)_3 \cdot 24H_2O$.

57. (c) $[Co(NH_3)_5NO_2]^{2+}$; $[Co(NH_3)_5ONO]^{2+}$
Two linkage isomers possible as NO_2 is an ambidentate ligand.

58. (a) $[MA_3B_3]$ type of compound exists are facial and meridonial isomer.

mer-isomer 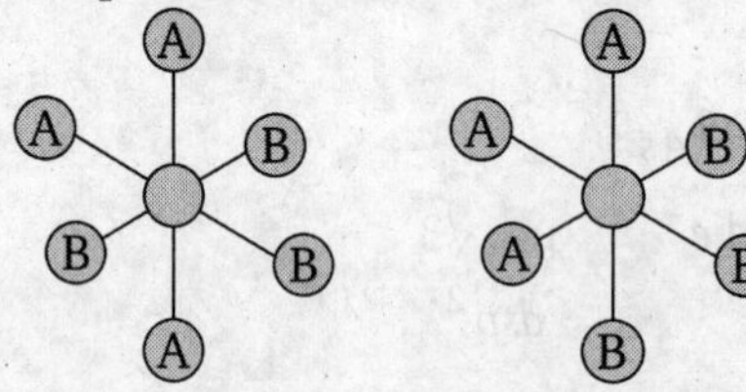 fac-isomer

59. (a) Smaller cations prefer coordinate bonding due to greater polarizing power.
Ti^{3+} = 67 pm radius, Cr^{3+} = 62 pm radius
Mn^{3+} = 65 pm radius, Fe^{3+} = 65 pm radius
So, Cr^{3+} has highest tendency to attract ligand.

60. (b) $[NiBr_2Cl_2]^{2-}$ → This complex species is tetrahedral as Br^- and Cl^- are weak field ligands.
$[PtBr_2Cl_2]^{2-}$ → As Pt belongs to 5d series. This complex species is square planar.
Both the complex species are optically inactive.
$[NiBr_2Cl_2]^{2-}$, being tetrahedral does not show Geometrical isomerism.
$[PtBr_2Cl_2]^{2-}$ shows two geometrical isomers.
Ni^{2+} and Pt^{2+} have different configurations. So different magnetic properties.

61. (b) A. $[Fe(CN)_6]^{3-}$; $n = 1$ (CN^- = Strong Field Ligand)
B. $[FeF_6]^{3-}$; $n = 5$ (F^- = Weak Field Ligand)
C. $[CoF_6]^{3-}$; $n = 4$ (F^- = WFL)
D. $[Cr(oxalate)_3]^{3-}$; $n = 3$ (oxalate = SFL)
E. $[Ni(CO)_4]$; $n = 0$ (CO = SFL)

62. (d) B. VBT does not explain stability of complex
C. Hybridisation of $[Ni(CN)_4]^{2-}$ is dsp^2.

63. (d) $Ni^{2+} + 2dmg^- \longrightarrow [Ni(dmg)_2]$

Rosy red/Bright Red precipitate.

64. (a) $\left.\begin{array}{l} K_2Cr_2O_7 \rightarrow Cr^{+6} \rightarrow \text{No } d-d \text{ transition} \\ KMnO_4 \rightarrow Mn^{7+} \rightarrow \text{No } d-d \text{ transition} \end{array}\right\}$ Charge transfer

65. (a) Mn, Ni and Cd metals are used in battery industries.

66. (c) 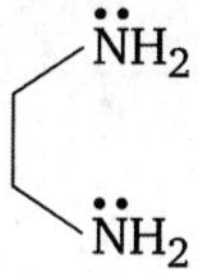 Bidentate, chelating

Based on Hall-Heroults process

$[Rh(PPh_3)_3Cl]$ Wilkinson's catalyst

$$As_2S + NaCN \underset{}{\overset{\text{Air}}{\rightleftharpoons}} Na[Ag(CN)_2] + Na_2S$$

Ca^{++} ion forms more stable complex with EDTA.

67. (a) $AlCl_3$ in acidified aqueous solution forms octahedral geometry $[Al(H_2O)_6]^{3+}$.

68. (d) For complex $K_3[Fe(SCN)_6]$

E.C. of Fe^{3+} – $[Ar]3d^5$ 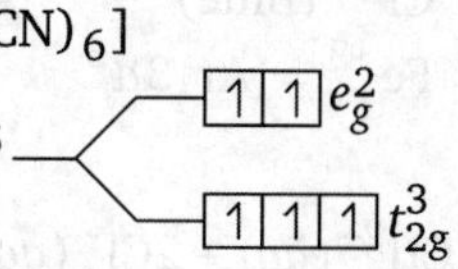

W.F.L.

Calculation of CFSE $= (-0.4 \times 3 + 0.6 \times 2)\,\Delta_0 = 0\,\Delta_0$

69. (b) $[NiCl_4]^{2-}$

$Ni^{2+} - [Ar]3d^8 4s^0 \rightarrow sp^3$, Tetrahedral

Number of unpaired electron = 2 paramagentic

$[Ni(CO)_4]$,

$Ni(0) \rightarrow [Ar]3d^{10} 4s^0$ (After rearrangement)

No unpaired electron

sp^3, Tetrahedral, Diamagnetic

70. (b) Order of CFSE

$[Co(NH_3)_4]^{2+} < [Co(NH_3)_6]^{2+} < [Co(NH_3)_6]^{3+} < [Co(en)_3]^{3+}$

Charge ↑ CFSE ↑ (between $[Co(NH_3)_6]^{2+}$ and $[Co(NH_3)_6]^{3+}$); On increasing strength of ligand CFSE ↑ (between $[Co(NH_3)_6]^{3+}$ and $[Co(en)_3]^{3+}$)

SFL : $NH_3 <$ en

71. (c) (A) $[Fe(CN)_5NO]^{2-} \rightarrow$ Heteroleptic, $Fe^{2+}, 3d^6, t_{2g}^6 e_g^0, d^2sp^3$, Low spin ($3d$ series + SFL)

(B) $[CoF_6]^{3-} \rightarrow$ Homoleptic, sp^3d^2, High spin, $Co^{3+}, 3d^6$ ($3d$ series + WFL)

(C) $[Fe(CN)_6]^{4-} \rightarrow$ Homoleptic $Fe^{2+}, 3d^6, d^2sp^3, t_{2g}^6 e_g^0$ Low spin ($3d$ series + SFL)

(D) $[Co(NH_3)_6]^{3+} \rightarrow$ Homoleptic, $CO^{3+}, 3d^6, d^2sp^3, t_{2g}^6 e_g^0$ Low spin ($3d$ series + SFL)

(E) $[Cr(H_2O)_6]^{2+} \rightarrow$ Homoleptic $Cr^{2+}, 3d^4, d^2sp^3$, High spin $t_{2g}^3 e_g^1$ ($3d$ series + WFL)

72. (c) $Co^{2+} = (Ar)_{18} 3d^7 4s^0$

1 1 e_g^2

⥮ ⥮ 1 t_{2g}^5

73. (a) Ma_3b_3 type complexes show Facial-Meridional isomerism

(i) $[Co(NH_3)_3Cl_3] \Rightarrow Ma_3b_3$

(ii) $[Co(NH_3)_4Cl_2]^+ \Rightarrow Ma_4b_2$

(iii) $[Co(en)_3]^{3+} \Rightarrow M(AA)_3$

(iv) $[Co(en)_2Cl_2]^+ \Rightarrow M(AA)_2b_2$

$a, b, = NH_3, Cl^-$

$AA = en$

74. (a) (a) V^{2+} (Violet), Cr^{3+} (Violet), Mn^{3+} (Violet)

(b) Zn^{2+} (Colourless), V^{3+} (Green), Fe^{3+} (Yellow)

(c) Ti^{4+} (Colourless), V^{4+} (Blue), Mn^{2+} (Pink)

(d) Sc^{3+} (Colourless), Ti^{3+} (Purple), Cr^{2+} (Blue)

75. (c) $Fe^{+6} = [Ar]3d^2$; $Mn^{+3} = [Ar]3d^4$; $Fe^{+3} = [Ar]3d^5$

$Cr^{+2} = [Ar]3d^4$; $Ni^{+4} = [Ar]3d^6$

76. (a) $[Co(NH_3)_5Cl]\,Cl_2 \longrightarrow \underbrace{[Co(NH_3)_5Cl]^{2+}(aq) + 2Cl^-(aq)}_{\text{3 ions in water}}$

$[Co(NH_3)_5Cl]Cl_2(aq) + 2AgNO_3(aq) \longrightarrow [Co(NH_3)_5Cl](NO_3)_2(aq) + 2AgCl(s)$

77. (b) $CoS \xrightarrow{\text{Aquaregia}} CoCl_2 \xrightarrow[\text{2. AcOH}]{\text{1. } KNO_2/NH_4OH} K_3[Co(NO_2)_6]\downarrow$ yellow

78. (c) $[Ni(H_2O)_6]^{2+}_{(aq)} + en_{(aq)} \rightarrow [Ni(H_2O)_4(en)]^{2+}_{(aq)} + 2H_2O$

Green

$[Ni(H_2O)_4(en)]^{2+}_{(aq)} + en_{(aq)} \rightarrow [Ni(H_2O)_2(en)_2]^{2+}_{(aq)} + 2H_2O$

Pale Blue

$[Ni(H_2O)_2(en)]^{2+}_{(aq)} + en_{(aq)} \rightarrow [Ni(en)_3]^{2+}_{(aq)} + 2H_2O$

Blue/purple

79. (a) For $3d^4$

If ligand is SFL : $t_{2g}^4\ e_g^0$ (Low spin)

If ligand is WFL : $t_{2g}^3\ e_g^1$ (High spin)

80. (c) (I) $[Mn(CN)_6]^{3-}$ $-1.6\,\Delta_o$

(II) $[Co(CN)_6]^{4-}$ $-1.8\,\Delta_o$

(III) $[Fe(CN)_6]^{4-}$ $-2.4\,\Delta_o$

(IV) $[Fe(CN)_6]^{3-}$ $-2.0\,\Delta_o$

I < II < IV < III

B. Objective Questions

[One or more than one option(s) is/are correct]

1. (c, d) $[Pt(en)_2Cl_2]Cl_2$

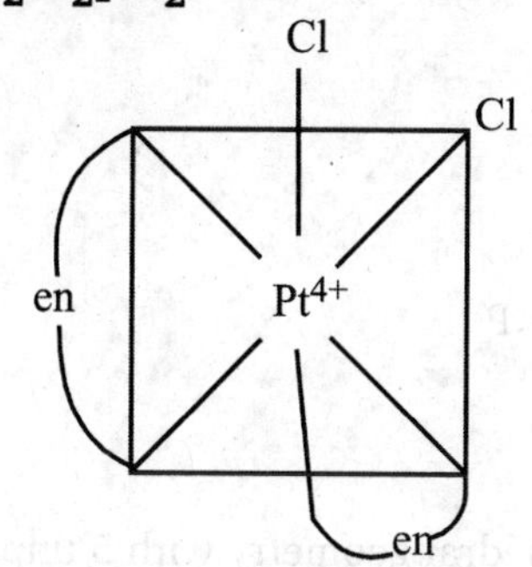

cis-form

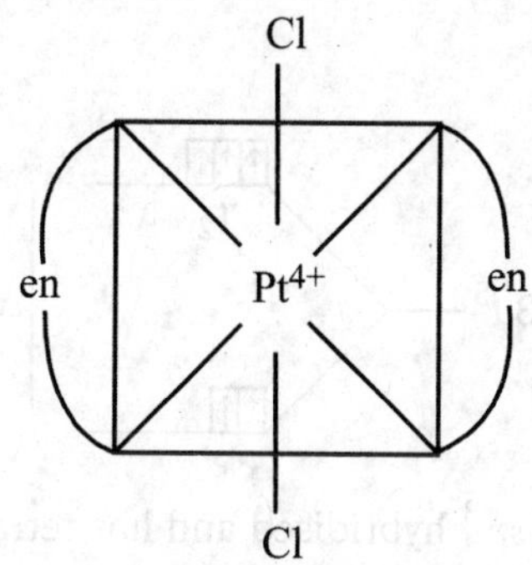

trans-form

$[Pt(NH_3)_2Cl_2]$

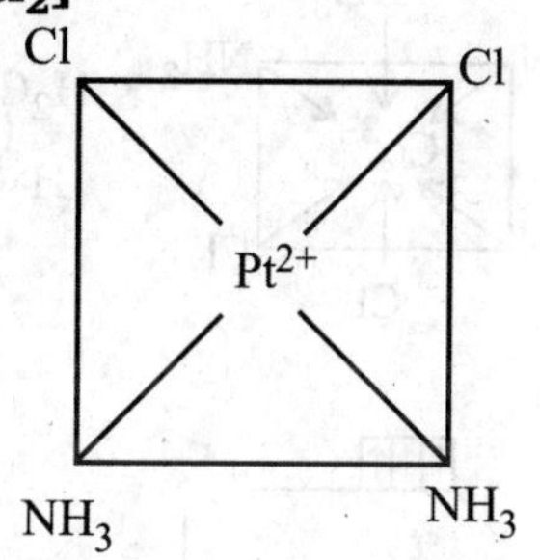

cis-form

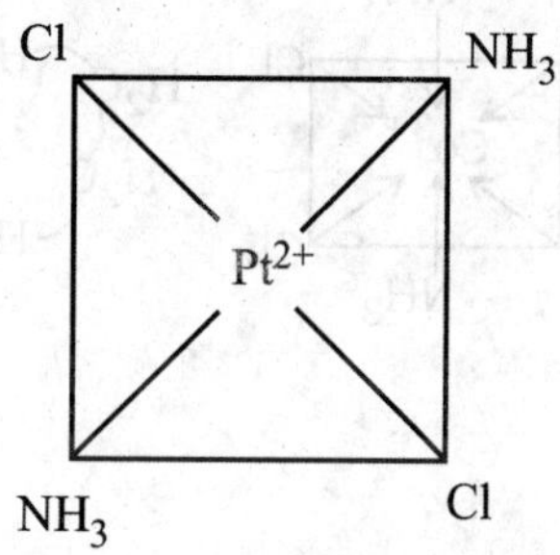

trans-form

2. (b, d)

(b) $[Co(NH_3)_4Cl_2]^2$ $\quad Ma_4b_2$

$[Pt(NH_3)_2(H_2O)Cl]^+$ $\quad Ma_2bc$

Both can show geometrical isomerism.

(d) $[Pt(NH_3)_2(NO_3)]Cl$

$[Pt(NH_3)_3Cl]Br$

Both can show ionization isomerism.

3. (a, b, c)

$$\underset{\substack{(X)\\ \text{Octahedral (pink)}\\ (\mu=3.87\text{ B.M.})}}{[Co(H_2O)_6]Cl_2} \xrightarrow[\text{air}]{\text{Aq. } NH_3 + NH_4Cl} \underset{\substack{(Y)\\ (\mu=0)}}{[Co(NH_3)_6]Cl_3}$$

$$\underset{(X)}{[Co(H_2O)_6]^{2+}} + HCl(\text{excess}) \overset{0^\circ C}{\rightleftharpoons} \underset{\substack{\text{tetrahedral (blue)}\\ (\mu=3.87\text{ B.M.})}}{[CoCl_4]^{2-}}, \quad \Delta H = +\text{ve}$$

4. (b, c)

(b) CO is powerful ligand so compounds having CO ligand are generally low spin complex.

(c) $M \underset{\sigma}{\overset{\pi}{\rightleftarrows}} C{\equiv}O$ Back bonding increase strength of metal and carbon bond. As metal get more $e^{\ominus}$ richer, back bonding get more stronger.

5. (a, c)

(a) $[FeCl_4]^-$

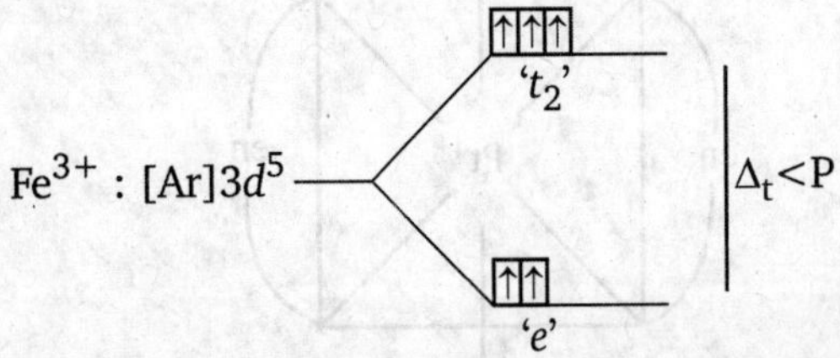

$[FeCl_4]^-$ is sp^3 hybridised and has tetrahedral geometry with 5 unpaired electrons.

(b) $[Co(en)(NH_3)_2Cl_2]^+$ has three geometrical isomers.

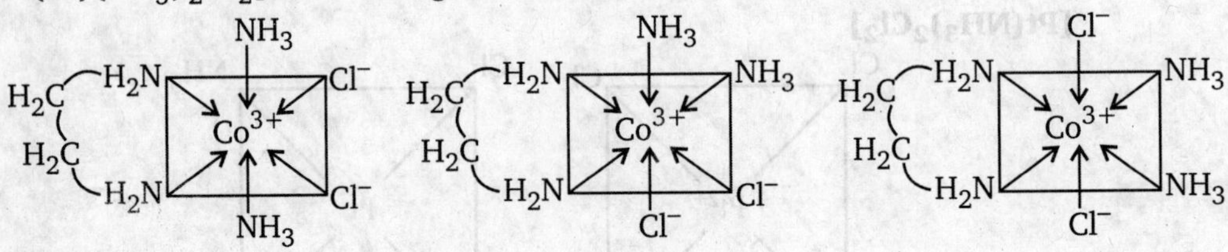

(c) $[FeCl_4]^-$

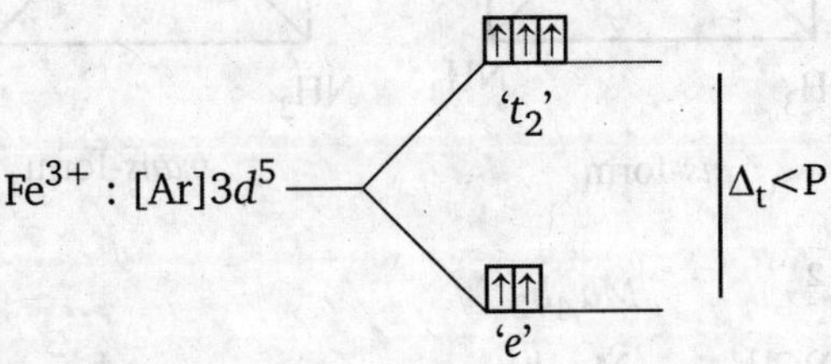

Number of unpaired electrons $(n) = 5$

Spin only magnetic moment $= \sqrt{n(n+2)}$ B.M.

$= 5.92$ B.M.

$[Co(en)(NH_3)_2Cl_2]^+$

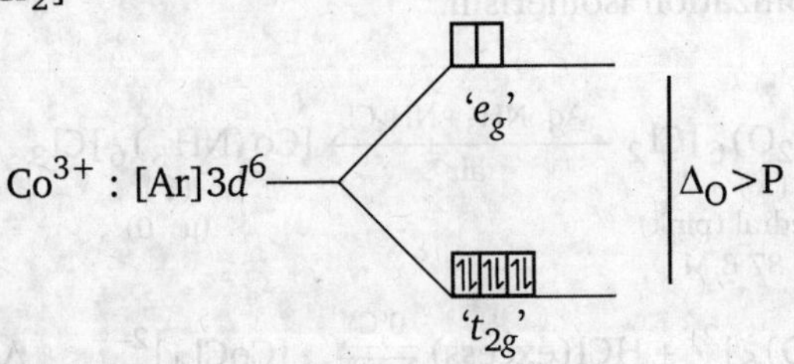

Number of unpaired electrons $(n) = 0$

Spin only magnetic moment $= \sqrt{n(n+2)}$ B.M.

(d) $[Co(en)(NH_3)_2Cl_2]^+$

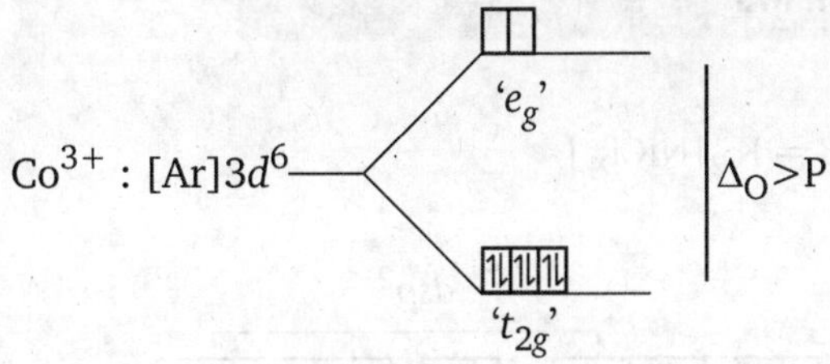

$[Co(en)(NH_3)_2Cl_2]^+$ is d^2sp^3 hybridised and has octahedral geometry with 0 unpaired electron.

C. Matching Type Problems

1. **a** – p, q, s; **b** – p, r, s; **c** – q, s; **d** – q, s

(a) $[Co(NH_3)_4Cl_2]Cl_2$: CO (II) : [Ar] $3d^7\ 4s^0$ ($n = 3$, paramagnetic)

$[Co(NH_3)_4.Cl_2]^{2+}$ is $M_{A_4B_2}$ type complex and exists in *cis* and *trans* form.

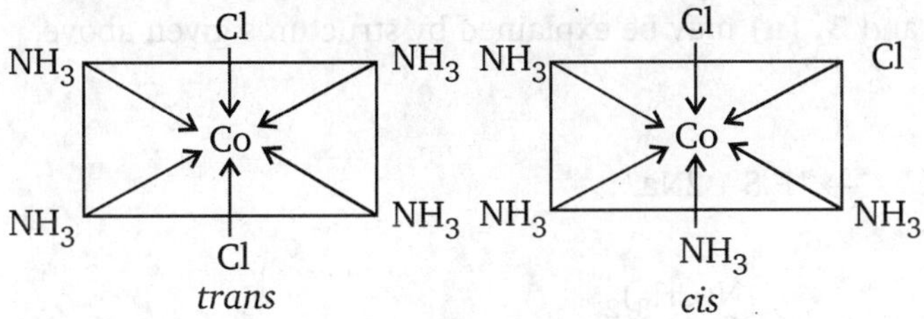

(b) $[Pt(NH_3)_2Cl_2]$:Pt (II) : [Xe] $5d^8, 6s^0$ ($n = 2$, paramagnetic)

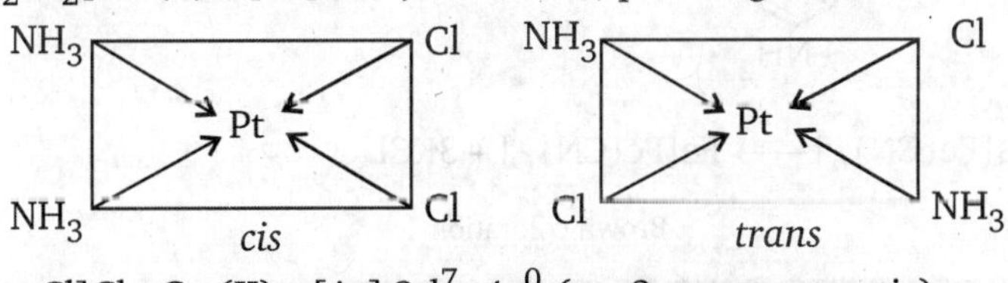

(c) $[Co(H_2O)_5.Cl]Cl$: Co (II) : [Ar] $3d^7, 4s^0$ ($n = 3$, paramagnetic).

M_{A_5B} type complex, so does not show geometrical isomerism.

(d) $[Ni(H_2O)_6]Cl_2$: Ni(II) : [Ar] $3d^8, 4s^0$ ($n = 2$, paramagnetic)

M_{A_6} type complex, so does not show geometrical isomerism.

2. (b)

(P) $[Cr(NH_3)_4Cl_2]Cl \longrightarrow$ Paramagnetic and exhibits *cis-trans* isomerism

(Q) $[Ti(H_2O)_5Cl](NO_3)_2 \longrightarrow$ Paramagnetic and exhibits ionization isomerism

(R) $[Pt(en)(NH_3)Cl]NO_3 \longrightarrow$ Diamagnetic and exhibits ionization isomerism

(S) $[Co(NH_3)_4(NO_3)_2]NO_3 \longrightarrow$ Diamagnetic and exhibits ionization *cis-trans* isomerism

D. Passage Based Problems

Passage-1

$A = K_2[Ni(CN)_4]$ $B = K_2[NiCl_4]$

$[Ni(CN)_4]^2$: Ni (II) :

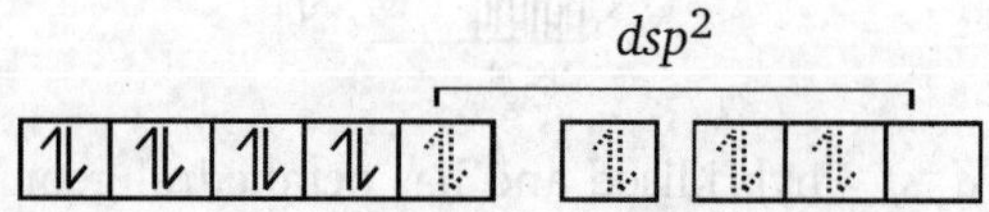

$n = 0$, diamagnetic.

$[NiCl_4]^{2-}$: Ni (II) :

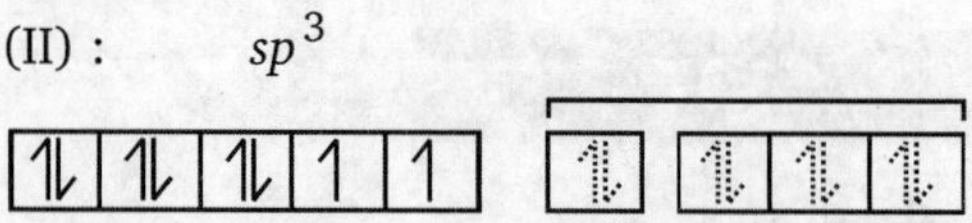

$n = 2$, paramagnetic

1. (a), 2. (c), and **3. (a)** may be explained by structures given above.

Passage-2

1. (d) $\underset{(X)}{Na_2S} + 2H^+ \longrightarrow H_2S + 2Na^+$

2. (c) 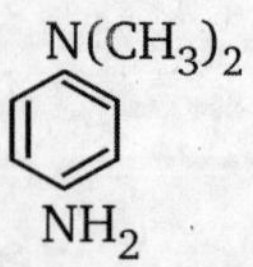 $+ H_2S +$ 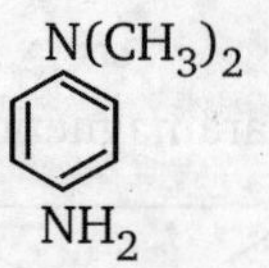 $+ \underset{(Y)}{6Fe^{3+}} \longrightarrow 6Fe^{2+} + NH_4^+ + 4H^+ + [\text{Methylene blue}]^+$

3. (b) $\underset{(Y)}{FeCl_3} + K_3[Fe(CN)_6] \longrightarrow \underset{\substack{(Z)\\ \text{Brown coloration}}}{Fe[Fe(CN)_6]} + 3KCl$

E. Statement and Explanation Type Problems

1. (a) $[Fe(H_2O)_5.NO]SO_4$: Fe (I) : [Ar] $3d^6$, $4s^1$

n (Theoretical value) = 5 (4 (3d) and 1 (4s) electrons)

This complex, $[Fe(H_2O)_5.NO]^{2+}$ is intensely coloured and intense colour is due to charge transfer spectra. Its magnetic moment is ≈3.9 B.M. which shows the presence of 3 unpaired electrons. It is explained by the fact that NO^+ leads to pairing of one 3d-electron by 4s electrons leaving 3 unpaired electrons.

Thus,

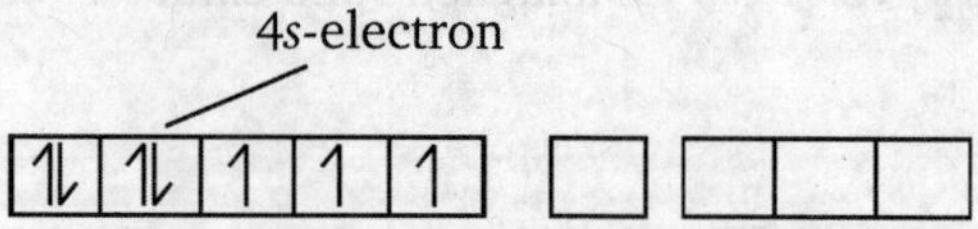

2. (a) The *cis* and *trans* both form of complex $[M(NH_3)_4Cl_2]$ are optically inactive.

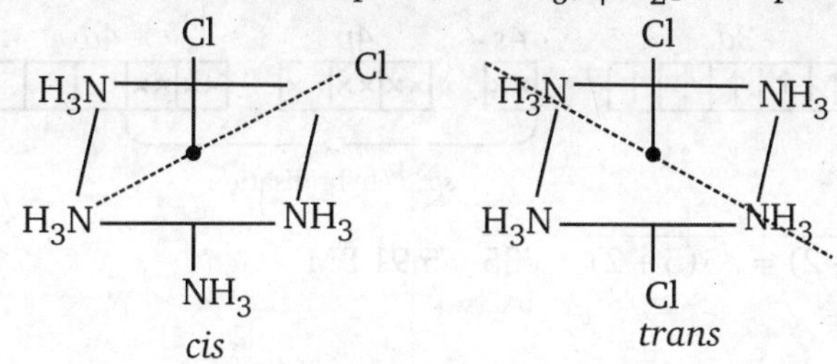

There are plane of symmetry in addition to that there is alternate axis of symmetry (C_n) which also causes their optical inactivity.

Note: If a molecule is not superimposable on its mirror image, that molecule can exhibit optical activity if these have proper axis but no improper axis or if a C_n axis be the only symmetry element present in a molecule, that molecule is optically active.

F. Integer Type Problems

1. (3) $[RhCl(CO)(PPh_3)(NH_3)]$ is m_{abcd} type square planar complex. Its three geometrical isomer are as given ahead.

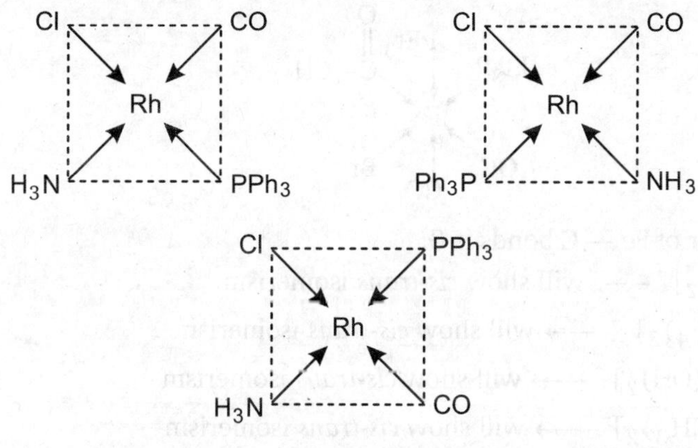

2. (8)

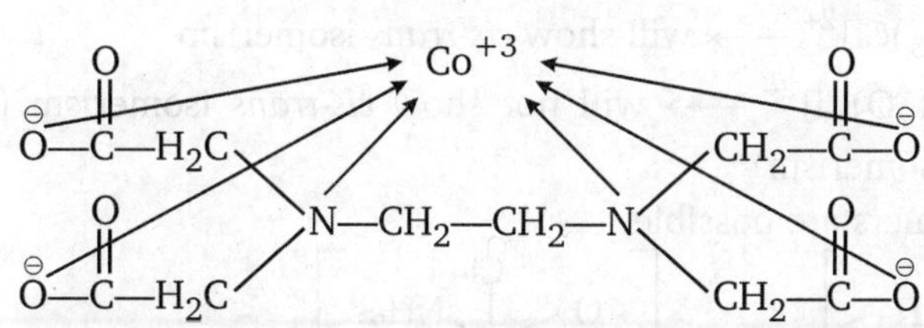

3. (4) $[Fe(SCN)_6]^{3-}$ and $[Fe(CN)_6]^{3-}$

In both the cases the electronic configuration of Fe^{3+} will be $1s^2, 2s^2, 2p^6, 3s^2, 3p^6, 3d^5$

In $[Fe(SCN)_6]^{3-}$, $\overline{S}CN$ is a weak field ligand and in $[Fe(CN)_6]^{3-}$, $\overline{C}N$ is a strong field ligand, the pairing will occur only in case of $[Fe(CN)_6]^{3-}$.

In $[Fe(SCN)_6]^{3-}$,

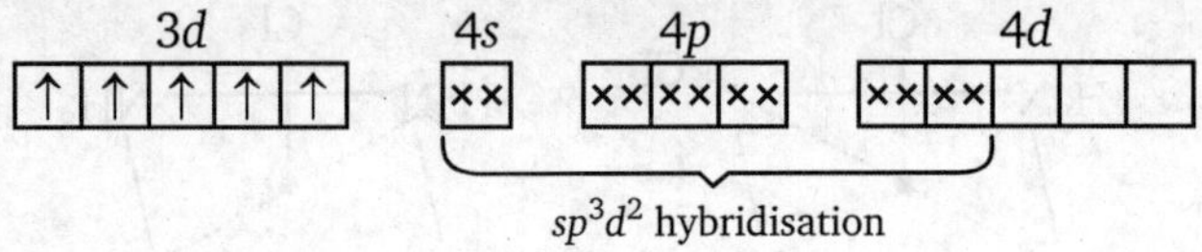

$\therefore \quad \mu = \sqrt{n(n+2)} = \sqrt{5(5+2)} = \sqrt{35} = 5.91$ BM

In $[Fe(CN)_6]^{3-}$,

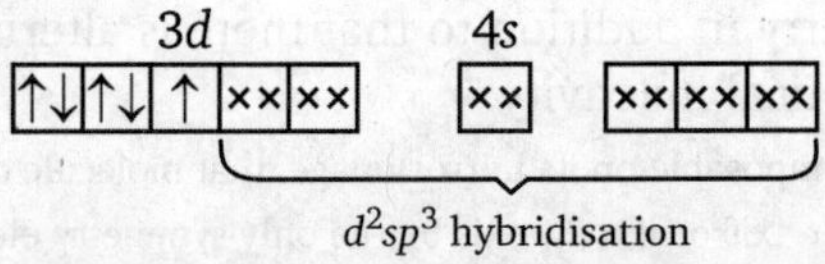

$$\mu = \sqrt{n(n+2)} = \sqrt{1(1+2)} = \sqrt{3} = 1.73 \text{ BM}$$

Difference in spin only magnetic moment = 5.91 – 1.73 = 4.18 ≈ 4 BM

4. (8) $[Fe(C_2O_4)(H_2O)]^{2-} + MnO_4^{2-} + 8H^+ \longrightarrow Mn^{2+} + Fe^{3+} + 4CO_2 + 6H_2O$

So the ratio of rate of change of $[H^+]$ to that of rate of change of $[MnO_4^-]$ is **8.**

5. (3) In complex, $[Fe(CH_3CO)(Br)(CO)_2(PEt_3)_2]$

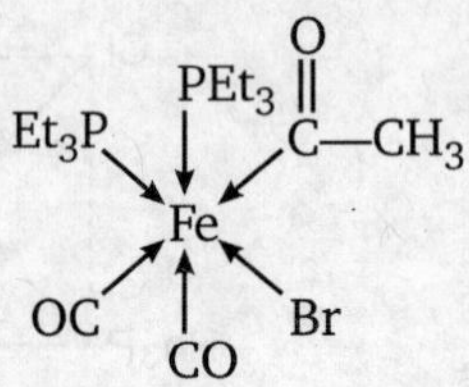

The number of Fe — C bonds is 3.

6. (6) $[Co(en)_2Cl_2]^+ \longrightarrow$ will show *cis-trans* isomerism

$[CrCl_2(C_2O_4)_2]^{3-} \longrightarrow$ will show *cis-trans* isomerism

$[Fe(H_2O)_4(OH)_2]^+ \longrightarrow$ will show *cis-trans* isomerism

$[Fe(CN)_4(NH_3)_2]^- \longrightarrow$ will show *cis-trans* isomerism

$[Co(en)_2(NH_3)Cl]^{2+} \longrightarrow$ will show *cis-trans* isomerism

$[Co(NH_3)_4(H_2O)Cl]^{2+} \longrightarrow$ will not show *cis-trans* isomerism (Although it will show geometrical isomerism)

7. (5) Total five isomers are possible.

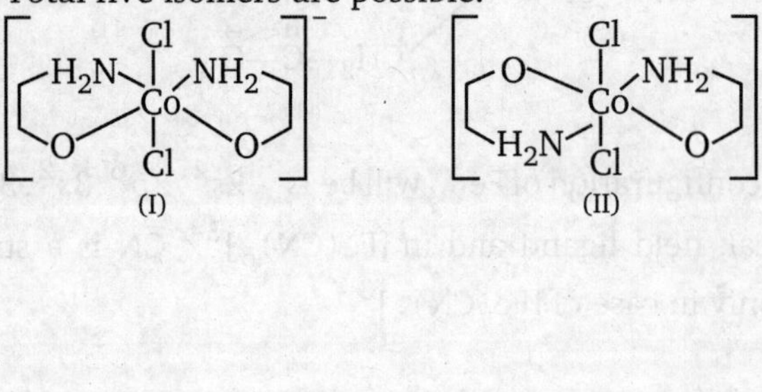

(I) (II)

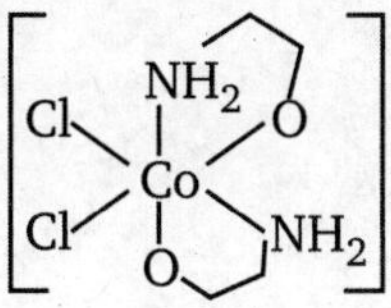

(III)

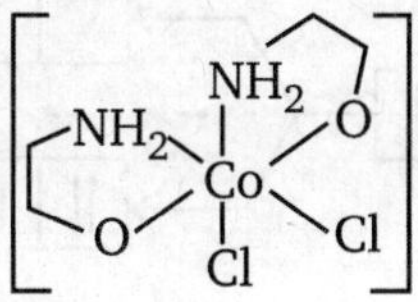

(IV)

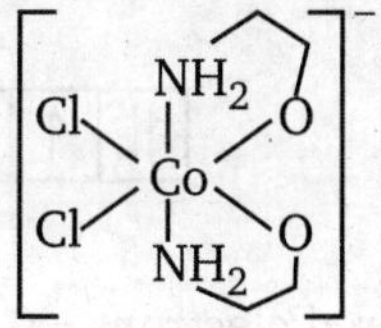

(V)

8. (1) Diamagnetic $\Rightarrow K_2\overset{+6}{Cr}O_4$ No unpaired electron

H atom $\Rightarrow$ 1 unpaired e^- $\Rightarrow$ paramagnetic

NO_2 monomer $\Rightarrow$ odd e^- species $\Rightarrow$ paramagnetic

$O_2^{\ominus}$ superoxide $\Rightarrow$ 1 unpaired e^- in π antibonding orbital

S_2 vapour $\Rightarrow$ like O_2 it contain 2 unpaired e^- in antibonding

$Mn_3O_4 \Rightarrow$ contain unpaired e^- (Mn^{2+}, Mn^{4+})

$[(NH_4)_2(FeCl_4)]$ or $(FeCl_4)^{2-} \Rightarrow Fe^{2+}$ with W.F.L. contain 4 unpaired e^-

$(NH_4)_2[NiCl_4] \Rightarrow \overset{+2}{Ni}$ with W.F.L. contain 2 unpaired e^-

9. (6) 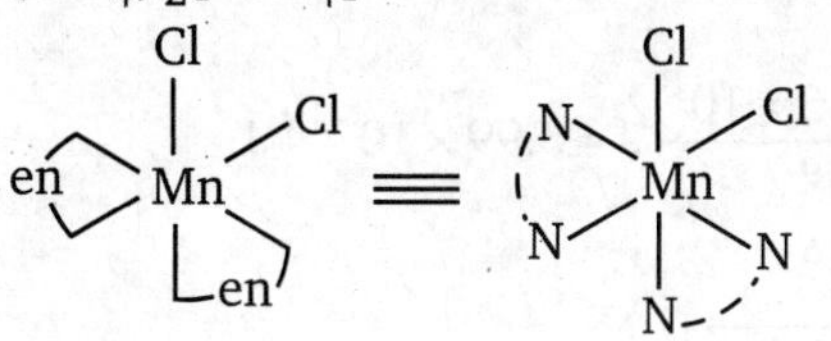

10. (26.92)

$[Co(NH_3)_6]Cl_3 + 3AgNO_3 \longrightarrow [Co(NH_3)_6]^{3+} + 3NO_3^- + 3AgCl(s)$

Moles of complex $= \dfrac{0.3}{267.46}$

Moles of $AgNO_3 = 3 \times \dfrac{0.3}{267.46} = \dfrac{0.125 \times V}{1000}$

$V = 26.92$ mL

11. (20)

Square pyramidal geometry Trigonal bipyramidal geometry

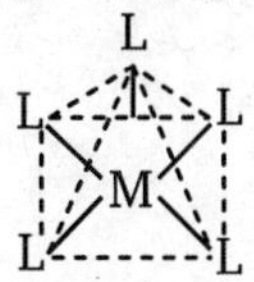

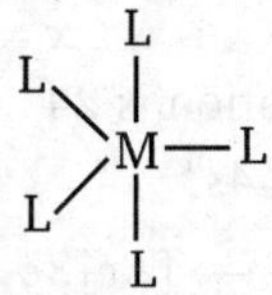

Square pyramidal	Trigonal bipyramidal
Number of 90° angles $\Rightarrow$ 8	Number of 90° angles $\Rightarrow$ 6
Number of 120° angles $\Rightarrow$ 0	Number of 120° angles $\Rightarrow$ 3
Number of 180° angles $\Rightarrow$ 2	Number of 180° angles $\Rightarrow$ 1

20 total

12. (0) $\Delta_0 > P$ pairing will take place.

$[Ru(H_2O)_6]^{2+}$ $\mu = \sqrt{n(n+2)}$ B.M.

$Ru^{2+} = [Kr]\,4d^6\,5s^0$

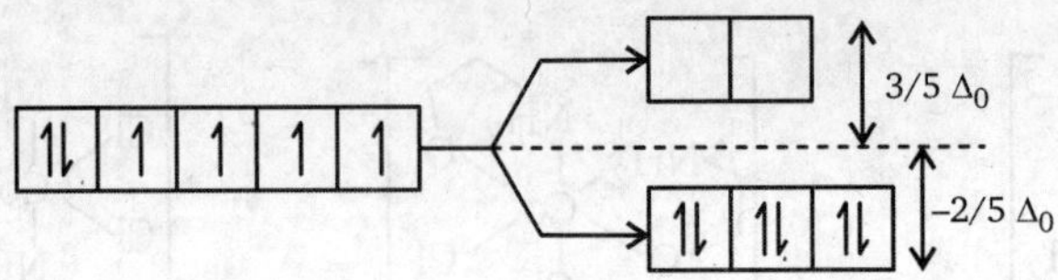

Unpaired electrons = 0

$$\mu = 0 \text{ B.M.}$$

13. (3) $[Co(ox)_2(Br)(NH_3)]^{2-}$

$[M(AA)_2bc]$ type of complex

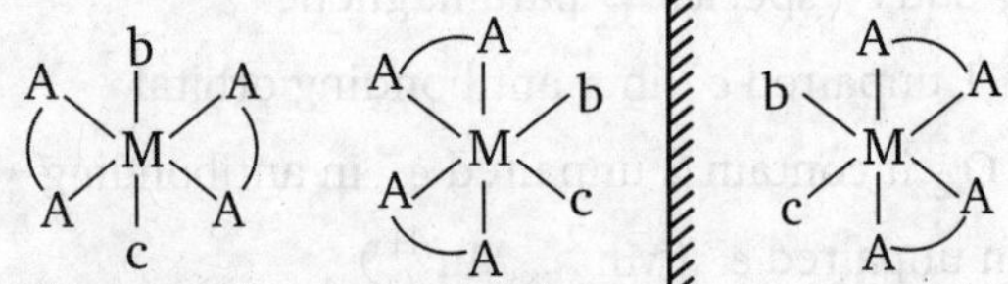

Optically inactive   Optically active

Total stereoisomers = 3

14. (4) $\lambda = 498 \text{ nm} = 498 \times 10^{-9}$ m

$$\text{Splitting energy } (E) = \frac{hc}{\lambda} = \frac{6.626 \times 10^{-34} \times 3 \times 10^{8}}{498 \times 10^{-9}} \text{ J} = 3.99 \times 10^{-19} \text{ J}$$

15. (49) Electronic configuration of Fe = $[Ar]4s^2 3d^6$

Unpaired electrons $(n) = 4$

Magnetic moment $(\mu) = \sqrt{n(n+2)} = \sqrt{4(4+2)}$

$$= \sqrt{24} = 4.89 \text{ B.M.} = 48.9 \times 10^{-1} \text{ B.M.}$$

16. (6) $FeCl_3 + 3KOH + 3H_2C_2O_4 \longrightarrow K_3[Fe(C_2O_4)_3] + 6H_2O + 3HCl$

Secondary valency will be $= 3 \times 2 \Rightarrow$ since each $C_2O_4^{2-}$ will acts as bidentate ligand.

17. (3) Let oxidation state of Cr is x.

$$3(+1) + x + 3(-2) = 0$$

$$x = +3$$

Cr electronic configuration is 24.

$$[Ar]3d^5 4s^1$$

$$Cr^{3+} \longrightarrow [Ar]3d^3$$

Unpaired electrons = 3

18. (9) Coordination no. = 6

Oxidation state is :

$$x - 4 = -1$$

$$x = +3$$

Sum = 3 + 6 = 9.

19. (3) The isomers of $[Cu(en)_2(SCN)_2]$ are :

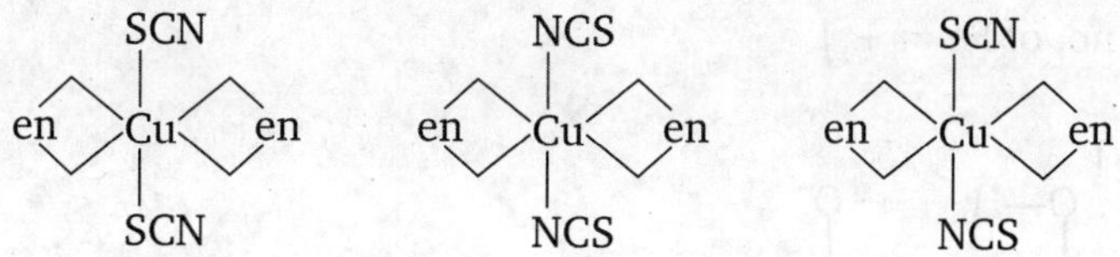

20. (3) For 1 : 1 electrolyte, the complex must be $[Co(NH_3)_4Cl_2]Cl$. Oxidation number of Co in $[Co(NH_3)_4Cl_2]^+$ is +3.

21. (6) Oxidation state of Mn

$x - 6 = -4$

$x = +2$

Br is weak field ligand

$$Mn^{2+} \longrightarrow t_{2g}^3 e_g^2;\ n = 5$$

$$\mu_s = \sqrt{n(n+2)} = \sqrt{35} = 5.91 = 6$$

22. (2) Fehling solution is a complex of Cu^{2+}

$Cu^{2+} = 3d^9$

No. of unpaired $e^- = 1$

$$\mu_S = \sqrt{1(1+2)} = \sqrt{3} = 1.73 \text{ BM}$$

23. (1) Spin magnetic moment of $[Cr(CN)_6]^{3-}$ $(t_{2g}^3 e_g^0)$

$\mu_1 = \sqrt{3(3+2)} = \sqrt{15}$ BM (Considering it as an outer orbital complex)

Spin magnetic moment of $[Cr(H_2O)_6]^{3+}$ $(t_{2g}^3 e_g^0)$

$$\mu_2 = \sqrt{3(3+2)} = \sqrt{15} \text{ BM}$$

$$\frac{\mu_1}{\mu_2} = \frac{\sqrt{15}}{\sqrt{15}} = 1$$

24. (3) $K_4[Fe(CN)_6] \Rightarrow Fe^{2+} = [Ar]3d^6$, CN $\Rightarrow$ SFL, $t_{2g}^6 e_g^0$

t_{2g} contain 6 electron so it become 3 pairs.

25. (4) $[Mn(NCS)_6]^{x-}$

Number of unpaired electron = 5

$\Rightarrow \quad 2 + (-6) = x \quad \Rightarrow \quad -4 = x \quad \Rightarrow \quad x = 4$

26. (480) $[Ti^{3+}(H_2O)_6]^{3+}$; $Ti^{3+} : 3d$

C.F.S.E. $= -0.4 \times \Delta_0 + 0$

$$\Rightarrow \quad -\frac{96 \times 10^3}{N_A} J = -0.4 \times \Delta_0 \quad \Rightarrow \quad \Delta_0 = \frac{96 \times 10^3}{0.4 \times 6 \times 10^{23}}$$

$$\Rightarrow \quad \frac{hc}{\lambda} = \frac{96 \times 10^3}{0.4 \times 6 \times 10^{23}}$$

$$\lambda = \frac{0.4 \times 6 \times 10^{23} \times 6.4 \times 10^{-34} \times 3 \times 10^8}{96 \times 10^3} = 0.48 \times 10^{-6} \text{ m} = 480 \times 10^{-9} \text{ m} = 480 \text{ nm}$$

27. (1) $[Fe(H_2O)_5(NO)]^{2-}$,

Oxidation no. of Fe = + 1

28. (14)

$[Ni(dmg)_2]$

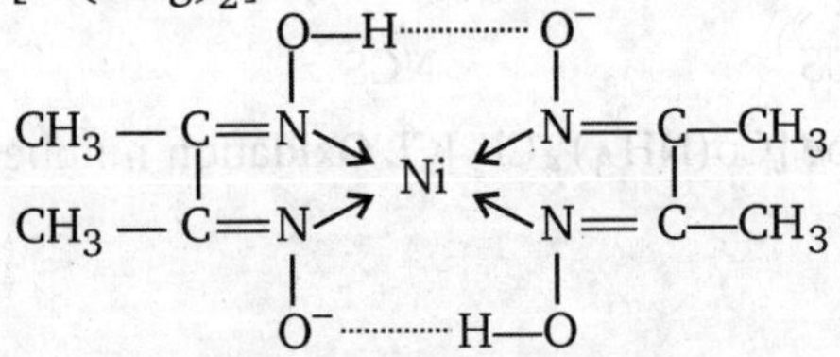

Number of H-atom = 14

29. (3) Only Cu^{2+}, Fe^{3+}, Ca^{2+} and Zn^{2+} form precipitate with $K_4[Fe(CN)_6]$.

30. (5) Strongest oxidising power among the option is Mn_2O_3 because of E^o value

$E^o_{Mn^{3+}/Mn^{2+}} = +1.57$ V

$Mn^{3+} \rightarrow d^4$ configuration

$\mu = \sqrt{24}$ BM = 4.89 BM ⇒ 5

31. (6) $O_2 \rightarrow 2$ unpaired electrons according to MOT

$O_2^+ \rightarrow 1$ unpaired electrons according to MOT

$O_2^- \rightarrow 1$ unpaired electrons according to MOT

$NO \rightarrow$ odd electron species

$NO_2 \rightarrow$ odd electron species

$K_2[NiCl_4] \rightarrow Ni^{2+} \Rightarrow 3d^8$ weak ligand, C.N. = 4

⇒ Tetrahedral, Paramagnetic with 2 unpaired electrons

32. (0) Yellow coloured and low spin complex are $K_3[Co(NO_2)_6]$ and $K_4[Fe(CN)_6]$.

(i) $K_3[Co(NO_2)_6]$

$Co^{3+} \Rightarrow [Ar]3d^6$

In presence of ligand field

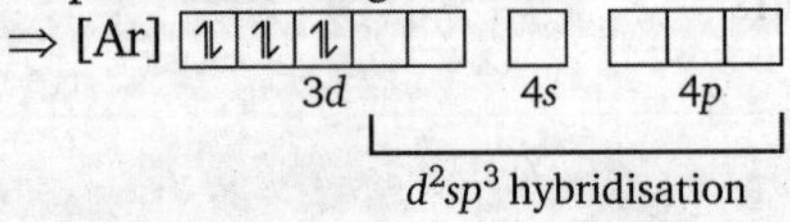

Magnetic moment = $\sqrt{n(n+2)}$ BM = 0 BM

(ii) $K_4[Fe(CN)_6]$

$Fe^{2+} \Rightarrow [Ar]3d^6$

In presence of ligand field

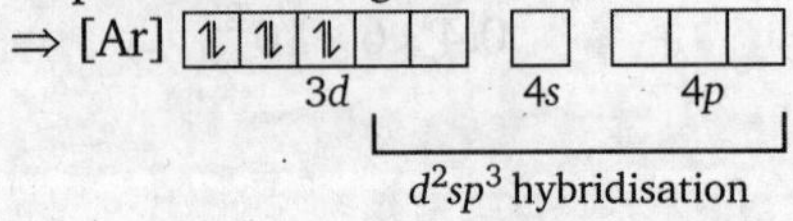

Magnetic moment = $\sqrt{n(n+2)}$ BM = 0 BM

❑❑❑

CHAPTER

9

Environmental Chemistry

A. Objective Questions

[Only one option is correct]

1. The gas leaked from a storage tank of the Union Carbide Plant in Bhopal gas tragedy was : **[JEE (Mains) 2013]**

(a) Methylisocyanate (b) Methylamine (c) Ammonia (d) Phosgene

2. The concentration of fluoride, lead, nitrate and iron in a water sample from an underground lake was found to be 1000 ppb, 40 ppb, 100 ppm and 0.2 ppm, respectively. This water is unsuitable for drinking due to high concentration of : **[JEE (Mains) 2016]**

(a) fluoride (b) lead (c) nitrate (d) iron

3. A water sample has ppm level concentration of following anions, $F^- = 10$; $SO_4^{2-} = 100$; $NO_3^- = 50$. The anion/anions that make/makes the water sample unsuitable for drinking is/are: **[JEE (Mains) 2017]**

(a) only F^- (b) only SO_4^{2-} (c) only NO_3^- (d) both SO_4^{2-} and NO_3^-

4. The recommended concentration of fluoride ion in drinking water is up to 1 ppm as fluoride ion is required to make teeth enamel harder by converting $[3Ca_3(PO_4)_2 \cdot Ca(OH)_2]$ to: **[JEE (Mains) 2018]**

(a) $[3Ca_3(PO_4)_2 \cdot CaF_2]$ (b) $[3\{Ca(OH)_2\} \cdot CaF_2]$

(c) $[CaF_2]$ (d) $[3(CaF_2) \cdot Ca(OH)_2]$

5. Which is wrong with respect to our responsibility as a human being to protect our environment? **[JEE (Mains) 2019]**

(a) Avoiding the use of floodlighted facilities

(b) Setting up compost tin in gardens

(c) Using plastic bags

(d) Restricting the use of vehicles

6. The maximum prescribed concentration of copper in drinking water is : **[JEE (Mains) 2019]**

(a) 0.05 ppm (b) 3 ppm (c) 5 ppm (d) 0.5 ppm

7. The layer of atmosphere between 10 km to 50 km above the sea level is called as: **[JEE Main 2019]**

(a) troposphere (b) thermosphere (c) stratosphere (d) mesosphere

8. The upper stratosphere consisting of the ozone layer protects us from the sun's radiation that falls in the wavelength region of: **[JEE Main 2019]**

(a) 600 – 750 nm (b) 0.8 – 1.5 nm (c) 400 – 550 nm (d) 200 – 315 nm

9. The regions of the atmosphere, where clouds form and where we live, respectively, are: **[JEE Main 2019]**

(a) stratosphere and stratosphere (b) stratosphere and troposphere
(c) troposphere and stratosphere (d) troposphere and troposphere

10. **Assertion:** Ozone is destroyed by CFCs in the upper stratosphere.
Reason: Ozone holes increase the amount of UV radiation reaching the earth. **[JEE Main 2019]**

(a) Assertion is false, but the reason is correct.
(b) Assertion and reason are incorrect.
(c) Assertion and reason are both correct, and the reason is the correct explanation for the assertion.
(d) Assertion and reason are correct, but the reason in not the explanation for the assertion.

11. The statement that is not true about ozone is: **[JEE Main (Sept.) 2020]**

(a) in the atmosphere, it is depleted by CFCs.
(b) in the stratosphere, it forms a protective shield against UV radiation.
(c) in the stratosphere, CFCs release chlorine free radicals (Cl) which reacts with O_3 to give chlorine dioxide radicals.
(d) it is a toxic gas and its reaction with NO gives NO_2

12. Thermal power plants can lead to: **[JEE Main (Sept.) 2020]**

(a) ozone layer depletion (b) acid rain
(c) blue baby syndrome (d) eutrophication

13. The processes of calcination and roasting in metallurgical industries, respectively, can lead to: **[JEE Main (Sept.) 2020]**

(a) photochemical smog and global warming
(b) global warming and photochemical smog
(c) global warming and acid rain
(d) photochemical smog and ozone layer depletion

14. Biochemical oxygen demand (BOD) is the amount of oxygen required (in ppm): **[JEE Main (Jan.) 2020]**

(a) for the photochemical breakdown of waste present in $1 m^3$ volume of a water body.
(b) for sustaining life in a water body.
(c) by bacteria to break-down organic waste in a certain volume of a water sample.
(d) by anaerobic bacteria to breakdown inorganic waste present in a water body.

15. The presence of soluble fluoride ion upto 1 ppm concentration in drinking water, is: **[JEE Main (Sept.) 2020]**

(a) harmful for teeth (b) harmful to skin (c) harmful to bones (d) safe for teeth

16. The incorrect statement(s) among (A)–(D) regarding acid rain is (are): **[JEE Main (Sept.) 2020]**

(A) It can corrode water pipes.
(B) It can damage structures made up of stone.
(C) It cannot cause respiratory ailments in animals
(D) It is not harmful for trees

(a) (A), (B) and (D) (b) (C) and (D) (c) (A), (C) and (D) (d) (C) only

17. Given below are two statements :

Statement I : Non-biodegradable wastes are generated by the thermal power plants.

Statement II : Biodegradable detergents leads to eutrophication.

In the light of the above statements, choose the most appropriate answer from the option given below : **[JEE Main (March) 2021]**

(a) Both statement I and statement II are false.
(b) Statement I is true but statement II is false.
(c) Statement I is false but statement II is true.
(d) Both statement I and statement II are true.

18. Given below are two statements : **[JEE Main (Feb.) 2021]**

Statement I : The pH of rain water is normally 5.6.

Statement II : If the pH of rain water drops below 5.6, it is called acid rain.

In the light of the above statements, choose the correct answer from the options given below.

(a) Statement I is true but Statement II is false.
(b) Both Statement I and Statement II are false.
(c) Statement I is false but Statement II is true.
(d) Both Statement I and Statement II are true.

19. Given below are two statements:

Statement I : The value of the parameter "Biochemical Oxygen Demand (BOD)" is important for survival of aquatic life.

Statement II : The optimum value of BOD is 6.5 ppm. In the light of the above statements, choose the most appropriate answer from the options given below: **[JEE Main (Feb.) 2021]**

(a) Statement I is false but Statement II is true.
(b) Both Statement I and Statement II are true.
(c) Statement I is true but Statement II is false.
(d) Both Statement I and Statement II are false.

20. The green house gas/es is (are) : **[JEE Main (March) 2021]**

(A) Carbon dioxide (B) Oxygen (C) Water vapour (D) Methane

Choose the most appropriate answer from the options give below:

(a) (A)and (C) only (b) (A) only
(c) (A), (C) and (D) only (d) (A) and (B) only

21. Which of the following statement (s) is (are) incorrect reason for eutrophication? **[JEE Main (March) 2021]**

(A) excess usage of fertilisers (B) excess usage of detergents
(C) dense plant population in water bodies

(D) lack of nutrients in water bodies that prevent plant growth

Choose the most appropriate answer from the options given below:

(a) (A) only (b) (C) only (c) (B) and (D) only (d) (D) only

22. The statements that are true: **[JEE Main (March) 2021]**

(A) Methane leads to both global warming and photochemical smog.

(B) Methane is generated from paddy fields.

(C) Methane is a stronger global warming gas than CO_2.

(D) Methane is a part of reducing smog.

Choose the most appropriate answer from the options given below:

(a) (A), (B), (C) only (b) (A) and (B) only

(c) (B), (C), (D) only (d) (A), (B), (D) only

23. The presence of ozone in troposphere: **[JEE Main (Feb.) 2021]**

(a) protects us from the UV radiation (b) protects us from the X-ray radiation

(c) protects us from greenhouse effect (d) generates photochemical smog

24. The type of pollution that gets increased during the day time and in the presence of O_3 is : **[JEE Main (March) 2021]**

(a) reducing smog (b) oxidising smog (c) global warming (d) acid rain

25. Reducing smog is a mixture of: **[JEE Main (March) 2021]**

(a) Smoke, fog and O_3 (b) Smoke, fog and SO_2

(c) Smoke, fog and $CH_2 = CH - CHO$ (d) Smoke, fog and N_2O_3

26. Given below are two statements:

Statement I : An allotrope of oxygen is an important intermediate in the formation of reducing smog.

Statement II : Gases such as oxides of nitrogen and sulphur present in troposphere contribute to the formation of photochemical smog. In the light of the above statements, choose the correct answer from the options given below. **[JEE Main (Feb.) 2021]**

(a) Both statement I and Statement II are false.

(b) Statement I is true but Statement II is false.

(c) Both Statement I and Statement II are true.

(d) Statement I is false but Statement II is true.

27. Which of the following are the Green house gases?

A. Water vapour B. Ozone

C. I_2 D. Molecular hydrogen

Choose the most appropriate answer from the options given. **[JEE Main (April) 2023 (II)]**

(a) B and C only (b) C and D only (c) A and D only (d) A and B only

28. The photochemical smog does not generally contain : **[JEE Main (July) 2022 (I)]**

(a) NO (b) NO_2 (c) SO_2 (d) HCHO

29. Sulphur dioxide is one of the components of polluted air. SO_2 is also a major contributor to acid rain. The correct and complete reaction to represent acid rain caused by SO_2 is : **[JEE Main (June) 2022 (II)]**

(a) $2SO_2 + O_2 \longrightarrow 2SO_3$ (b) $SO_2 + O_3 \longrightarrow SO_3 + O_2$

(c) $SO_2 + H_2O_2 \longrightarrow H_2SO_4$ (d) $2SO_2 + O_2 + 2H_2O \longrightarrow 2H_2SO_4$

30. Polar stratospheric clouds facilitate the formation of : **[JEE Main (June) 2022 (I)]**
(a) $ClONO_2$ (b) HOCl (c) ClO (d) CH_4

31. The eutrophication of water body results in : **[JEE Main (June) 2022 (I)]**
(a) loss of biodiversity (b) breakdown of organic matter
(c) increase in biodiversity (d) decrease in BOD

32. Which one of the following elemental forms is not present in the enamel of the teeth? **[JEE Main (June) 2022 (I)]**
(a) Ca^{2+} (b) P^{3+} (c) F^- (d) P^{5+}

33. The radical which mainly causes ozone depletion in the presence of UV radiations is : **[JEE Main (April) 2023 (I)]**
(a) $\dot{C}H_3$ (b) $NO^\bullet$ (c) $Cl^\bullet$ (d) $\dot{O}H$

34. The delicate balance of CO_2 and O_2 is NOT disturbed by : **[JEE Main (April) 2023 (II)]**
(a) Burning of coal (b) Deforestation
(c) Burning of petroleum (d) Respiration

35. The industrial activity held least responsible for global warming is : **[JEE Main (Feb.) 2023 (II)]**
(a) Steel manufacturing
(b) Electricity generation in thermal power plants
(c) Manufacturing of cement
(d) Industrial production of urea

36. The possibility of photochemical smog formation is more at : **[JEE Main (April) 2023 (I)]**
(a) The places with healthy vegetation (b) Himalayan villages in winter
(c) Marshy lands (d) Industrial area

37. The groups of chemicals used as pesticide is **[JEE Main (April) 2023 (II)]**
(a) Sodium chlorate, DDT, PAN (b) Aldrin, Sodium chlorate, Sodium arsinite
(c) DDT, Aldrin (d) Dieldrin, Sodium arsinite, Tetrachloroethene

38. The water quality of a pond was analysed and its BOD was found to be 4. The pond has : **[JEE Main (Jan.) 2023 (II)]**
(a) Highly polluted water
(b) Water has high amount of fluoride compounds
(c) Very clean water
(d) Slightly polluted water

B. Statement and Explanation Type Problems

Read the following questions and answer as per the direction given below :
(a) Assertion and reason are correct, but the reason is not the explanation for the assertion.
(b) Assertion is false, but the reason is correct.
(c) Assertion and reason are incorrect.
(d) Assertion and reason are both correct and the reason is the correct explanation for the assertion.

1. Assertion : Ozone is destroyed by CFCs in the upper stratosphere.

Reason : Ozone holes increase the amount of UV radiation reaching the earth.

[JEE (Mains) 2019]

Answers

[A] Objective Questions (Only one option is correct)

1. (a)	**2.** (c)	**3.** (a)	**4.** (a)	**5.** (c)	**6.** (b)	**7.** (c)	**8.** (d)	**9.** (d)	**10.** (d)
11. (c)	**12.** (b)	**13.** (c)	**14.** (c)	**15.** (d)	**16.** (b)	**17.** (d)	**18.** (d)	**19.** (c)	**20.** (c)
21. (d)	**22.** (a)	**23.** (a)	**24.** (b)	**25.** (b)	**26.** (b)	**27.** (d)	**28.** (c)	**29.** (d)	**30.** (b)
31. (a)	**32.** (b)	**33.** (c)	**34.** (d)	**35.** (d)	**36.** (d)	**37.** (c)	**38.** (c)		

[B] Statement and Explanation Type Problems

1. (a)

Hints and Solutions

A. Objective Questions

[Only one option is correct]

1. (a) Factual.

2. (c) Fluoride, lead and iron are present within their permissible limits but nitrate ion which has permissible value of 50 ppm, is present in much higher amount *i.e.*, 100 ppm which makes the water unfit for drinking.

3. (a) Above 500 ppm of SO_4^{2-} ions in drinking water, can cause laxative effect otherwise lesser ppm value is permissible for drinking.

Maximum limit of NO_3^- ions in drinking water is 50 ppm, above this limit it can cause the disease like methemoglobinemia.

More than 1 ppm F^- ions in drinking water are not fit for drinking, it can cause decay of bones and teeth.

4. (a) The F^- ions make the enamel on teeth much harder by converting hydroxyapatite. $[3Ca(PO_4)_2 \cdot Ca(OH)_2]$, into much harder fluorapatite *i.e.*, $[3Ca(PO_4)_2 \cdot CaF_2]$

7. (c) Layer of atmosphere between 10 km to 50 km above sea level is called stratosphere .

8. (d) Ozone layer protects us from UV rays which range from 200-315 nm.

9. (d) Troposphere is region where living organisms is found and cloud forms.

10. (d) CFC reacts with ozone and decompose it, results in ozone layer depletion. Ozone layer protects us from harmful UV rays.

Depletion in ozone layer increases the amount of UV radiation.

11. (c) $CF_2Cl_2 \xrightarrow{UV} \dot{Cl}(g) + \dot{C}F_2Cl(g)$

$\dot{Cl}(g) + O_3(g) \longrightarrow Cl\dot{O}(g) + O_2(g)$

Chlorine radical ($\dot{Cl}$) reacts with O_3 to give chlorine oxide ($Cl\dot{O}$) radical.

12. (b) Thermal power plants produces large amount of nitrogen oxides and sulphur dioxides that cause acid rain.

13. (c) During calcination and roasting process, CO_2 and SO_2 gases are released. These gases caused global warming and are responsible for acid rain.

14. (c) Biochemical oxygen demand (BOD) is amount of oxygen required by bacteria to breakdown organic waste in certain volume of water sample.

15. (d) Presence of soluble fluoride ion upto 1 ppm concentration in drinking water is safe for teeth to prevent tooth decay.

16. (b) Acid rain causes respiratory issue in animals and Human being. Acid rain removes minerals and nutrients from oils that trees need to grow. So, acid rain also harmful for trees.

17. (d) Non-biodegradable wastes are generated by the thermal power plants which produce fly ash.

Biodegradable detergents are known to be toxic, corrosive and contribute to growing levels of eutrophication in water bodies.

18. (d) Rain water is acidic due to dissolution of CO_2 in water.

$$H_2O + CO_2 \longrightarrow H_2CO_3$$

$$H_2CO_3(aq) \rightleftharpoons H^+(aq) + HCO_3^-(aq)$$

Normally, pH of rain water is 5.6. If pH of rain water drops below 5.6 it is called acid rain. So, both statements are correct.

19. (c) Statement (I) is correct. Statement (II) is incorrect. Since, optimum value of BOD is 5 ppm.

20. (c) Green house gases are CO_2, CH_4 and water vapours.

21. (d) In eutrophication water have excess of nutrients and support plant growth.

22. (a) Methane is a major greenhouse gas that causes both global warming and photochemical smog. It is mainly generated by paddy fields.

CH_4 is 40 times stronger greenhouse gas than CO_2. Methane is not reducing smog. Reducing smog is a mixture of smoke, fog and SO_2.

23. (a) Presence of ozone in troposphere protect earth from ultraviolet ray.

24. (b) Ozone is a common oxidizing agent found in photochemical smog. It increases in day time due to automobiles and factories exhaust.

25. (b) Reducing smog is a mixture of smoke, fog and SO_2.

26. (b) $NO_2(g) \xrightarrow{h\nu} NO(g) + O(g)$; $\quad O_2(g) + O(g) \rightleftharpoons O_3(g)$

Ozone (O_3) is important intermediate in the formation of reducing smog so, statement (I) is correct. Photochemical smogs consists of ozone, nitrogen dioxide and peroxyacetylnitrate (PAN). Statement (II) is incorrect.

27. (d) Green house gases are CO_2, CH_4, water vapour, nitrous oxide, CFCs and ozone.

28. (c) Photochemical smog contains nitrogen dioxide (NO_2), Ozone (O_3), PAN (peroxyacetylnitrate), and compounds containing —CHO group.

29. (d) SO_2 is atmospherically oxidized in presence of water to give sulphuric acid

$$SO_2 + O_2 + 2H_2O \longrightarrow 2H_2SO_4 \text{ (Acid rain)}$$

30. (b) Polar stratospheric clouds provide surface on which hydrolysis of $ClONO_2$ takes place to form HOCl (Hypochlorous acid)

$$ClONO_2(g) + H_2O(g) \longrightarrow HOCl(g) + HNO_3(g)$$

31. (a) Nutrient enriched water bodies support a dense plant population which can deprive oxygen from animal life. This results in biodiversity.

32. (b) Calcium (Ca^{2+}) and phosphate (PO_4^{3-}) are the major components of teeth enamel.

33. (c)

$$O_2(g) \xrightarrow{UV} O(g) + O(g)$$

$$O_2(g) + O(g) \longrightarrow O_3(g)$$

$$CF_2Cl_2(g) \xrightarrow{UV} \dot{Cl}(g) + \dot{CF_2Cl}(g)$$

$$\dot{Cl}(g) + O_3(g) \longrightarrow \dot{ClO}(g) + O_2(g)$$

$$\dot{ClO}(g) + O(g) \longrightarrow \dot{Cl}(g) + O_2(g)$$

34. (d) Respiration, is a natural process. So balance of CO_2 and O_2 is not disturbed by respiration.

35. (d) In urea production NH_3 and CO_2 consumed so least responsible for global warming.

36. (d) Photochemical smog occurs in warm, dry and sunny climate. The main components come from the action of sunlight on unsaturated hydrocarbon and nitrogen oxides produced by automobiles and factories.

37. (c) D.D.T. and Aldrin are used as pesticides.

38. (c) Clean water has BOD value of <5 while polluted water has BOD of 15 or more.

□□□

CHAPTER

10

Salt Analysis

A. Objective Questions

1. The compound formed in the positive test for nitrogen with the Lassaigne solution of an organic compound is : **[AIEEE 2004]**

(a) $Fe_4[Fe(CN)_6]_3$ (b) $Na_3[Fe(CN)_6]$

(c) $Fe(CN)_3$ (d) $Na_4[Fe(CN)_5]NOS$

2. Biuret test is not given by : **[AIEEE 2010]**

(a) Proteins (b) Carbohydrates

(c) Polypeptide (d) Urea

3. Which of the following reagents may be used to distinguish between phenol and benzoic acid ? **[AIEEE 2011]**

(a) Aqueous NaOH (b) Tollen's reagent

(c) Molisch reagent (d) Neutral $FeCl_3$

4. Passing H_2S gas into a mixture of Mn^{2+}, Ni^{2+}, Cu^{2+} and Hg^{2+} ions in an acidified aqueous solution precipitates : **[JEE Advanced 2011]**

(a) CuS and HgS (b) MnS and CuS

(c) MnS and NiS (d) NiS and HgS

5. Upon treatment with ammoniacal H_2S, the metal ion that precipitates as a sulfide is : **[JEE Advanced 2012]**

(a) Fe (III) (b) AI (III) (c) Mg (II) (d) Zn (II)

6. Among PbS, CuS, HgS, MnS, Ag_2S, NiS, CoS, Bi_2S_3 and SnS_2, the total number of black coloured sulphides is : **[JEE Advanced 2014]**

(a) 4 (b) 6 (c) 5 (d) 3

7. The pair (s) of ions where both the ions are precipitated upon passing H_2S gas in presence of dilute HCl, is (are) : **[JEE Advanced 2015]**

(a) Ba^{2+}, Zn^{2+} (b) Bi^{3+}, Fe^{3+}

(c) Cu^{2+}, Pb^{2+} (d) Hg^{2+}, Bi^{2+}

8. The hottest region of Bunsen flame shown in the figure below is : **[JEE Main 2016]**

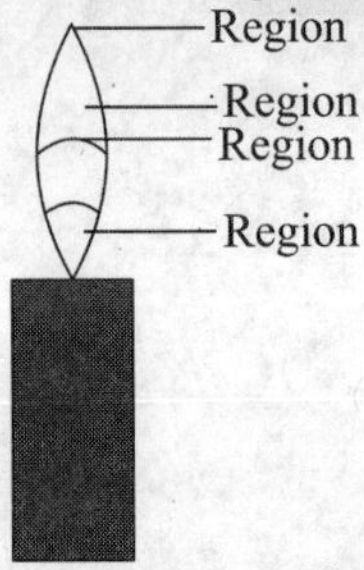

(a) Region 1 (b) Region 2
(c) Region 3 (d) Region 4

9. The reagent(s) that can selectivity precipitate S^{2-} and SO_4^{2-} in aqueous solution is (are) : **[JEE Advanced 2016]**

(a) $CuCl_2$ (b) $BaCl_2$
(c) $Pb(OOCCH_3)_2$ (d) $Na_2[Fe(CN)_5NO]$

10. In the following reaction sequence in aqueous solution, the species *X*, *Y* and *Z* respectively,

$$S_2O_3^{2-} \xrightarrow{Ag^+} \underset{\text{Clear solution}}{X} \xrightarrow{Ag^+} \underset{\text{White precipitate}}{Y} \xrightarrow{\text{Withtime}} \underset{\text{Black precipitate}}{Z}$$ **[JEE Advanced 2016]**

(a) $[Ag(S_2O_3)_2]^{3-}$, $Ag_2S_2O_3$, Ag_2S (b) $[Ag(S_2O_3)_3]^{5-}$, Ag_2SO_3, Ag_2S
(c) $[Ag(SO_3)_2]^{3-}$, $Ag_2S_2O_3$, Ag (d) $[Ag(SO_3)_2]^{3-}$, Ag_2SO_4, Ag

11. Solution containing a group – IV cation gives a precipitate on passing H_2S. A solution of this precipitate in dil. HCl produces a white precipitate solution and bluish white precipitate with basic potassium ferro cyanide. The cation is : **[JEE Main 2017]**

(a) Co^{2+} (b) Ni^{2+}
(c) Zn^{2+} (d) Mn^{2+}

12. For standardizing NaOH solution, which of the following is used as a primary standard? **[JEE Main 2018]**

(a) Oxalic acid (b) Ferrous ammonium sulphate
(c) Sodium tetraborate (d) Dil. HCl

13. An alkali is titrated against an acid with methyl orange as indicator, which of the following is a correct combination? **[JEE Main 2018]**

	Base	Acid	End point
(a)	Weak	Strong	Yellow to pinkish-red
(b)	Strong	Strong	Pink to colourless
(c)	Weak	Strong	Colourless to pink
(d)	Strong	Strong	Pinkish-red to yellow

14. A chemist has 4 samples of artificial sweetener *A*, *B*, *C* and *D*. To identify these samples, he performed certain experiments and noted the following observations :

(i) *A* and *D* both form blue-violet colour with ninhydrin.

(ii) Lassaigne extract of *C* gives positive $AgNO_3$ test and negative $Fe_4[Fe(CN)_6]_3$ test.

(iii) Lassaigne extract of *B* and *D* gives positive sodium nitroprusside test.
Based on these observations which option is correct? **[JEE Main (Jan.) 2020]**
(a) *A* : aspartame ; *B* : saccharin ; *C* : sucralose ; *D* : alitame
(b) *A* : alitame ; *B* : saccharin ; *C* : aspartame ; *D* : sucralose
(c) *A* : saccharin ; *B* : alitame ; *C* : sucralose ; *D* : aspartame
(d) *A* : aspartame ; *B* : alitame ; *C* : Saccharin ; *D* : sucralose

15. The incorrect statement is : **[JEE Main (Sep.) 2020]**
(a) In manganate and permanganate ions, the *p*-bonding takes place by overlap of *p*-orbitals of oxygen and *d*-orbitals of manganese.
(b) Manganate ion is green in colour and permanganate ion in purple in colour.
(c) Manganate and permanganate ions are paramagnetic.
(d) Manganate and permanganate ions are tetrahedral.

16. The addition of dilute NaOH to Cr^{3+} salt solution will give : **[JEE Main 2021]**
(a) A solution of $[Cr(OH)_4]^-$ (b) Precipitate of $[Cr(OH)_6]^{3-}$
(c) Precipitate of $Cr_2O_3(H_2O)_n$ (d) Precipitate of $Cr(OH)_3$

17. Match List-I with List-II.

List-I (Metal Ion)		List-II (Group in Qualitative Analysis)	
(A)	Mn^{2+}	(i)	Group-III
(B)	As^{3+}	(ii)	Group-IIA
(C)	Cu^{2+}	(iii)	Group-IV
(D)	Al^{3+}	(iv)	Group-IIB

Choose the most appropriate answer from the options given below : **[JEE Main 2021]**
(a) (A)-(iv), (B)-(ii), (C)-(iii), (D)-(i)
(b) (A)-(iii), (B)-(iv), (C)-(ii), (D)-(i)
(c) (A)-(i), (B)-(iv), (C)-(ii), (D)-(iii)
(d) (A)-(i), (B)-(ii), (C)-(iii), (D)-(iv)

18. Given below are two statements: One is labelled as Assertion A and the other is labelled as Reason R. **[JEE Main 2022]**
Assertion A: Permanganate titrations are not performed in presence of hydrochloric acid.
Reason R: Chlorine is formed as a consequence of oxidation of hydrochloric acid.
In the light of the above statements, choose the correct answer from the options given below.
(a) Both A and R are true and R is the correct explanation of A
(b) Both A and R are true but R is not the correct explanation of A
(c) A is true but R is false
(d) A is false but R is true

19. A white precipitate was formed when $BaCl_2$ was added to water extract of an inorganic salt. Further, a gas 'X' with characteristic odour was released when the formed white precipitate was dissolved in dilute HCl. The anion present in the inorganic salt is : **[JEE Main 2022]**

(a) I^- (b) SO_3^{2-} (c) S^{2-} (d) NO_2^-

20. The reaction of $Pb(NO_3)_2$ and NaCl in water produces a precipitate that dissolves upon the addition of HCl of appropriate concentration. The dissolution of the precipitate is due to formation of : **[JEE Advanced 2022]**

(a) $PbCl_2$ (b) $PbCl_4$ (c) $[PbCl_4]^{2-}$ (d) $[PbCl_6]^{2-}$

21. During the qualitative analysis of SO_3^{2-} using dilute H_2SO_4, SO_2 gas evolved which turns $K_2Cr_2O_7$ solution (acidified with dilute H_2SO_4) : **[JEE Main 2023]**

(a) red (b) black
(c) blue (d) green

22. In the wet tests for identification of various cations by precipitation, which transition element cation doesn't belong to group IV in qualitative inorganic analysis ? **[JEE Main 2023]**

(a) Co^{2+} (b) Zn^{2+} (c) Ni^{2+} (d) Fe^{3+}

23. Match List-I with List-II.

List-I (Complexes)		**List-II (Hybridisation)**	
(A)	$[Ni(CO)_4]$	(I)	sp^3
(B)	$[Cu(NH_3)_4]^{2+}$	(II)	dsp^2
(C)	$[Fe(NH_3)_6]^{2+}$	(III)	sp^3d^2
(D)	$[Fe(H_2O)_6]^{2+}$	(IV)	d^2sp^3

Choose the most appropriate answer from the options given below : **[JEE Main 2023]**

(a) A-II, B-I, C-IV, D-III (b) A-I, B-II, C-IV, D-III
(c) A-I, B-II, C-III, D-IV (d) A-II, B-I, C-III, D-IV

24. Formulae of Nessler's reagent is : **[JEE Main 2023]**

(a) K_2HgI_4 (b) HgI_2 (c) KHg_2I_2 (d) $KHgI_3$

25. Given below are two statements:

Statement I : Dimethyl glyoxime form a six-membered covalent chelate when treated with $NiCl_2$ solution in presence of NH_4OH.

Statement II : Prussian blue precipitate contains iron both in (+2) and (+3) oxidation states.

In the light of the above statements, choose the most appropriate answer from the options given below: **[JEE Main 2024]**

(a) Statement I is false but Statement II is true.
(b) Both Statement I and Statement II are true.
(c) Both Statement I and Statement II are false.
(d) Statement I is true but Statement II is false.

26. The metal ion whose electronic configuration is not affected by the nature of the ligand and which gives a violet colour in non-luminous flame under hot condition is borax bead test is : **[JEE Main (Jan.) 2025]**

(a) Ti^{3+} (b) Ni^{2+} (c) Mn^{2+} (d) Cr^{3+}

27. Identify the inorganic sulphides that are yellow in colour :

(A) $(NH_4)_2S$ (B) PbS (C) CuS (D) As_2S_3

(E) As_2S_5

Choose the correct answer from the options given below : **[JEE Main (Jan.) 2025]**

(a) (A) and (C) only (b) (A), (D) and (E) only

(c) (A) and (B) only (d) (D) and (E) only

B. Passage Based Problems

Passage-1

An aqueous solution of a mixture of two inorganic salts, when treated with dilute HCl, gave a precipitate (*P*) and a filtrate (*Q*). The precipitate (*P*) was found to dissolve in hot water. The filtrate (*Q*) remained unchanged, when treated with H_2S in a dilute mineral acid medium. However, it gave a precipitate (*R*) with H_2S in an ammoniacal medium. The precipitate (*R*) gave a coloured solution (*S*), when treated with H_2O_2 in an aqueous NaOH medium.

1. The coloured solution (*S*) contains : **[JEE Advanced 2013]**

(a) $Fe_2(SO_4)_3$ (b) $CuSO_4$

(c) $ZnSO_4$ (d) Na_2CrO_4

2. The precipitate (*P*) contains : **[JEE Advanced 2013]**

(a) Pb^{2+} (b) Hg_2^{2+}

(c) Ag^+ (d) Hg^{2+}

C. Integer Answer Type Problems

1. Number of metal ions characterized by flame test among the following is ______.

$Sr^{2+}, Ba^{2+}, Ca^{2+}, Cu^{2+}, Zn^{2+}, Co^{2+}, Fe^{2+}$ **[JEE Main 2024]**

Answers

[A] Objective Questions (Only one option is correct)

1. (a)	**2.** (b)	**3.** (d)	**4.** (a)	**5.** (d)	**6.** (b)	**7.** (c,d)	**8.** (b)	**9.** (a)	**10.** (a)
11. (c)	**12.** (a)	**13.** (a)	**14.** (a)	**15.** (c)	**16.** (d)	**17.** (b)	**18.** (a)	**19.** (b)	**20.** (d)
21. (d)	**22.** (d)	**23.** (b)	**24.** (a)	**25.** (a)	**26.** (b)	**27.** (d)			

[B] Passage Based Problems

Passage-1

1. (d) **2.** (a)

[C] Integer Type Problems

1. (4)

Hints and Solutions

A. Objective Questions

1. (a) $3Na_4[Fe(CN)_6] + 4Fe_3 \longrightarrow \underset{\text{Prussian blue}}{Fe_4[Fe(CN)_6]_3} + 12Na^+$

2. (b) Biuret test is used to characterize the presence of —CONH group in a compound.

3. (d) Phenol gives violet colouration with neutral ferric chloride solution.
Benzoic acid gives buff coloured (pale dull yellow) precipitate with neutral ferric chloride solution.

4. (a) H_2S in presence of aqueous acidified solution precipitates as sulphide of Cu and apart from Pb^{+2}, Bi^{+3}, Cd^{+2}, As^{+3}, Sb^{+3} and Sn^{+3}.

5. (d) H_2S gas in presence of NH_4Cl and NH_4OH is the group reagent in group IVth radicals *i.e.*, Zn^{2+}, Ni^{2+}, Mn^{2+} and Co^{2+}.

6. (b) The total number of black coloured sulfides is 6. These include PbS, CuS, HgS, NiS, CoS, Bi_2S_3.
Bi_2S_3 in its crystalline form is dark brown but Bi_2S_3 precipitate obtained is black in colour. Hence, the correct option is (b).

7. (c,d) The pairs of ions where both the ions are precipitated upon passing H_2S gas in presence of dilute HCl, are Cu^{2+}, Pb^{2+} and Hg^{2+}, Bi^{2+}. The ions Cu^{2+}, Pb^{2+}, Hg^{2+} and Bi^{2+} belong to group IIA. They precipitate as sulphides. The group reagent is hydrogen sulphide in presence of HCl. HCl suppresses the ionization of hydrogen sulphide (due to common ion effect) so that only solubility product of sulphides of II group radicals is overpowered. Hence, the radicals of other groups (II, IV and V) do not precipitate.

8. (b) Region-2 is the blue flame which is the hottest region of Bunsen flame.

9. (a) (a) Cu^{2+} will give black precipitate of CuS while $CuSO_4$ is soluble.

$$Cu^{2+} + SO_4^{2-} \longrightarrow \underset{\text{Black precipitate}}{CuS\downarrow}$$

$$Cu^{2+} + SO_4^{2-} \longrightarrow \underset{\text{Soluble}}{CuSO_4}$$

(b) Ba^{2+} will give white precipitate of $BaSO_4$ while BaS is soluble.

$$Ba^{2+} + S^{2-} \longrightarrow \underset{\text{Soluble}}{BaS}$$

$$Ba^{2+} + SO_4^{2-} \longrightarrow \underset{\text{White precipitate}}{BaSO_4}$$

(c) Pb^{2+} will give precipitate with both S^{2-} and SO_4^{2-}.

$$Pb^{2+} + SO_4^{2-} \longrightarrow \underset{\text{Black precipitate}}{PbS\downarrow}$$

$$Pb^{2+} + SO_4^{2-} \longrightarrow \underset{\text{White precipitate}}{PbSO_4\downarrow}$$

(d) $$2Na^+ + S^{2-} \longrightarrow Na_2S \xrightarrow{Na_2[Fe(CN)_5NO]} \underset{\text{Purple solution}}{Na_2[Fe(CN)_5NOS]}$$

$$2Na^+ + S_4^{2-} \longrightarrow \underset{\text{Soluble}}{Na_2SO_4}$$

10. (a) $$\underset{\text{(Excess)}}{Ag^+ + 2S_2O_3^{2-}} \longrightarrow \underset{\substack{\text{Argento thiosulphate } (X) \\ \text{(Soluble)}}}{[Ag(S_2O_3)_2]^{3-}} \xrightarrow{Ag}$$

$$\underset{(Y)\text{ (White ppt.)}}{Ag_2S_2O_3} \xrightarrow[\text{(Due to hydrolysis)}]{\text{Withtime}} \underset{(Z)\text{ (Black ppt.)}}{Ag_2S}$$

11. (c) $$Zn^{2+} \longrightarrow ZnS$$
$$\downarrow \text{dil. HCl}$$
$$\underset{\text{(White ppt.)}}{Zn(OH)_2} \xleftarrow{NaOH} ZnCl_2 \xrightarrow{K_4[Fe(CN)_6]} \underset{\text{(Bluish white ppt.)}}{Zn_2[Fe(CN)_6]}$$

12. (a) Oxalic acid is used as a primary standard for standardizing sodium hydroxide solution.

13. (a) Base — Weak, Acid — Strong, End point — Yellow to pinkish-red

Methyl orange shows pinkish-red colour in acidic medium and yellow colour in the basic medium. Since the original solution is basic so the initial colour is yellow and titrated with acid so the final colour is pinkish-red.

Therefore option (a) is correct.

14. (a) (i) Blue-violet colour with ninhydrin → amino acid derivative. So it cannot be saccharide or sucralose.

(ii) Lassaigne extract give +ve test with $AgNO_3$. So Cl is present, –ve test with $Fe_4[Fe(CN)_6]_3$ means N is absent. So it can't be aspartame or saccharine or alitame, so *C* is sucralose.

(iii) Lassaigne solution of *B* and *D* given +ve sodium nitroprusside test, so it is having S, so it is saccharine and alitame.

(a) Aspartame

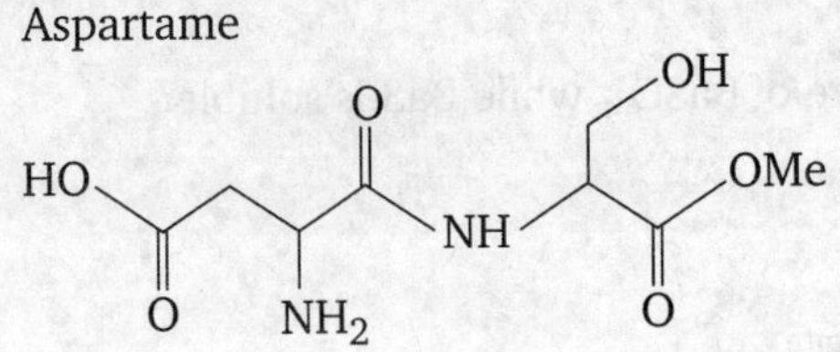

(b) Saccharine

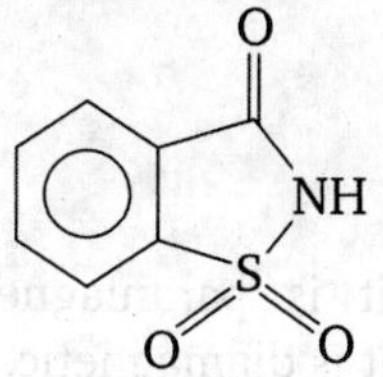

(c) Sucralose

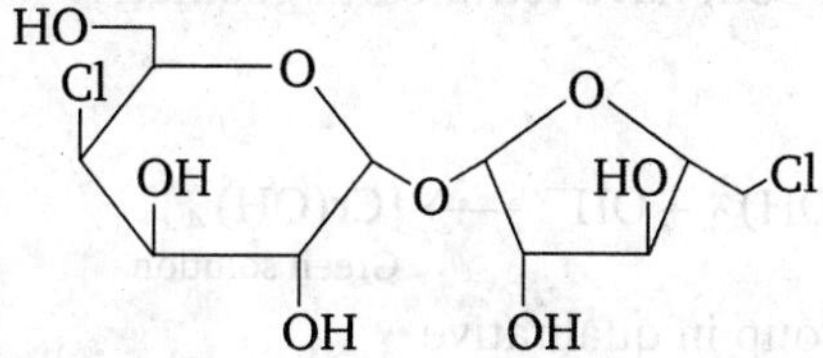

(d) Alitame

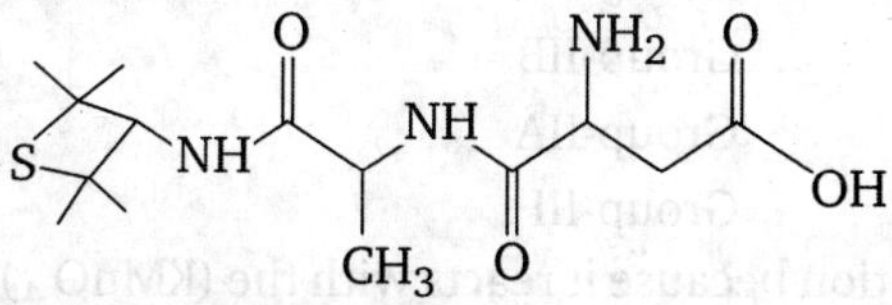

15. (c) (a) Manganate $\Rightarrow MnO_4^{2-}$, Permanganate $\Rightarrow MnO_4^-$

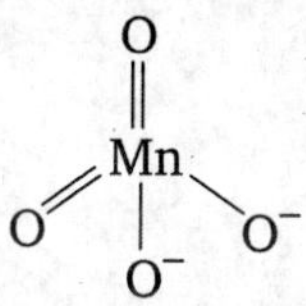

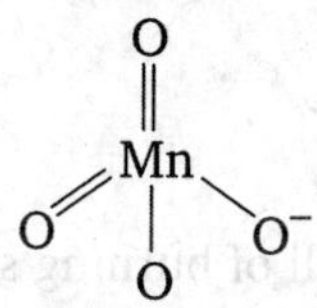

Hybridisation of Mn $\Rightarrow d^3s$, Hybridisation of Mn $\Rightarrow d^3s$

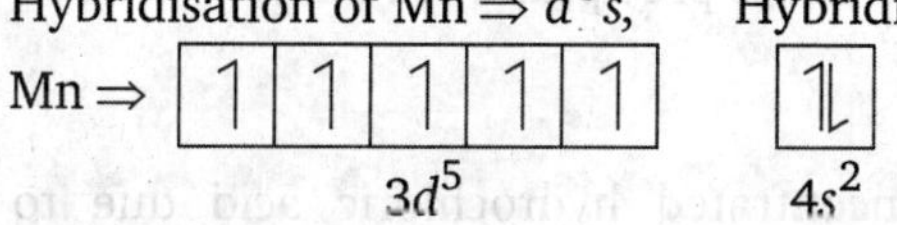

After excitation

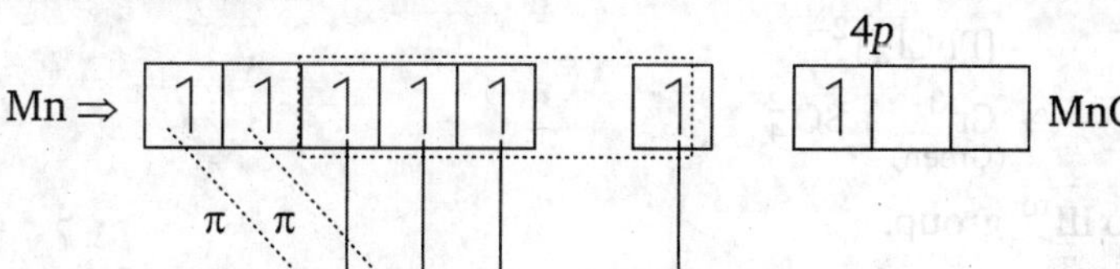

$2 \times 2p_\pi - 3d_{\pi\sigma}$

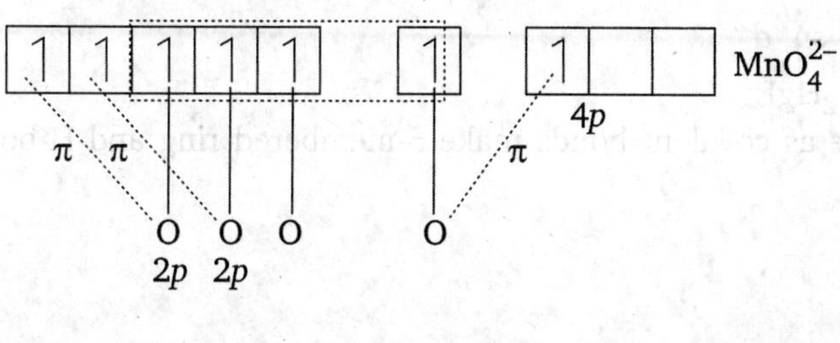

$2 \times 2p_\pi - 3d_\pi$

$1 \times 2p_\pi - 4p_\pi$

(b) $MnO_4^{2-} \Rightarrow$ green

$MnO_4^- \Rightarrow$ purple/violet

(c) Manganate contains 1 unpaired electron hence it is paramagnetic where as permanganate contains no unpaired electrons hence it is diamagnetic.

(d) Both have d^3s hybridisation hence both have tetrahedral geometry.

16. (d) $Cr^{3+} + OH^- \longrightarrow \underset{\text{(Green ppt.)}}{Cr(OH)_3}$

If NaOH is present in excess, then $Cr(OH)_3 + OH^- \longrightarrow \underset{\text{Green solution}}{[Cr(OH)_4]^-}$

17. (b)

Metal ion analysis	Group in qualitative
Mn^{2+}	Group-IV
As^{3+}	Group-IIB
Cu^{2+}	Group-IIA
Al^{3+}	Group-III

18. (a) HCl is not used in the process of titration because it reacts with the ($KMnO_4$) that is used in the process and gets oxidized.

19. (b) Anion is SO_3^{2-}

$BaSO_3 \xrightarrow{\text{dil. HCl}} \underset{X\text{(gas)}}{SO_2 \uparrow}$

Gas is released with smell of burning sulphur.

20. (d) Pb^{2+} on reaction with Cl^-, produces white precipitate of $PbCl_2$

$Pb^{2+} + 2Cl^- \longrightarrow \underset{\text{White}}{PbCl_2\downarrow}$

This precipitate is soluble in concentrated hydrochloric acid due to formation of tetrachloroplumbate (II) ion

$PbCl_2 \downarrow + 2Cl^- \longrightarrow [PbCl_4]^{2-}$

21. (d) $SO_2 + Cr_2O_7^{2-} \longrightarrow \underset{\text{(Green)}}{Cr^{3+}} + SO_4^{2-}$

22. (d) Fe^{3+} belongs to IIIrd group.

23. (b) $Ni(CO)_4 \longrightarrow sp^3$

$[Cu(NH_3)_4]^{2+} \longrightarrow dsp^2$

$[Fe(NH_3)_6]^{2+} \longrightarrow d^2sp^3$

$[Fe(H_2O)_6]^{2+} \longrightarrow sp^3d^2$

24. (a) Nessler's reagent K_2HgI_4

25. (a) Statement I is false as covalent bonds make 5-membered ring and H-bonding makes 6 membered ring.

$Ni^{2+} + NH_4OH + dmg \longrightarrow$ 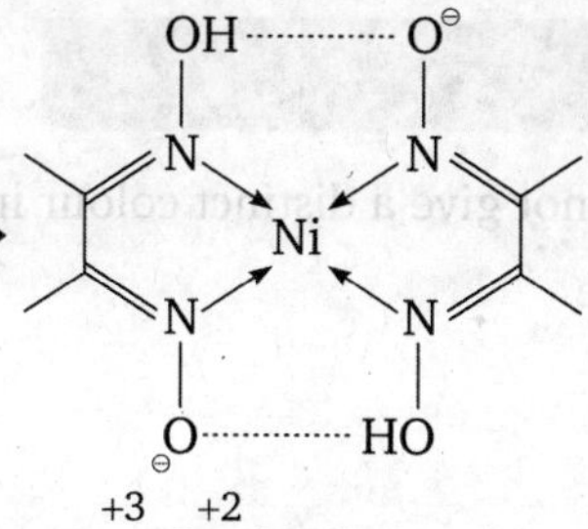

$\equiv \overset{+3}{Fe_4}\,[\overset{+2}{Fe}(CN)_6]_3$

26. (b) Ni^{2+} gives violet coloured bead in non-luminous flame under hot conditions. Ni^{2+} has d^8 configuration which does not depend on nature of ligand present in octahedral complex.
$Ni^{2+}: t_{2g}^6 e_g^2$

27. (d) As_2S_3 and As_2S_5 are yellow colour sulphides, $(NH_4)_2S$ is colourless, PbS is black, CuS is black in colour.

B. Passage Based Problems

Passage-1

1. (d)

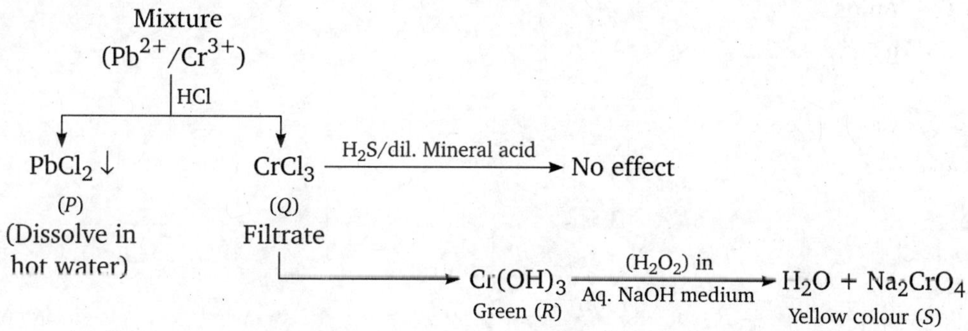

2. (a)

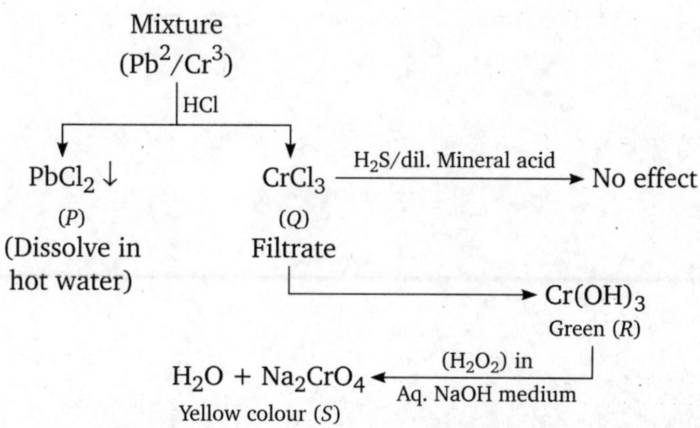

C. Integer Type Problems

1. (4) Zn^{2+}, Co^{2+} and Fe^{2+} do not give a distinct colour in the flame test.

Cu^{2+} ⇒ Blue-green flame

Ca^{2+} ⇒ Orange-red flame

Ba^{2+} ⇒ Green flame

Sr^{2+} ⇒ Red flame

❑❑❑

Note

Note

Note

Note